Theory of Strict Convexity, Best Approximation *and* Fixed Points

Dr Meenu Sharma

PARTRIDGE
A Penguin Random House Company

ISBN: Softcover   978-1-4828-5649-1
               eBook   978-1-4828-5648-4

Print information available on the last page.

To order additional copies of this book, contact
Partridge India
000 800 10062 62
orders.india@partridgepublishing.com

www.partridgepublishing.com/india

CONTENTS

INTRODUCTION

The present book is a study of strictly convex linear metric spaces and some results on Approximation and Fixed Points in Metric linear spaces and Metric spaces. In this introductory chapter we give a brief historical background of the subject (Strict Convexity, Approximation Theory applications of fixed point theorems to approximation theory)and a chapter wise summary of the results contained in this book.

The notion of strict convexity in normed linear spaces was introduced independently by J.A. Clarkson and M.G. krein in 1935 (see[11]). Different mathematicians calls strictly convex normed spaces by different names such as strictly normalized , rotund and strongly convex (see[11]). Ahuja , Naran and Trehan ([1]) extended the notion of strict convexity to linear metric spaces in 1977 . Later the study was pursued by K.P.R. Sastry, S.V.R. Naidu, T.D. Narang, concept or strict convexity to linear metric spaces, whatever literature was available, the uniquencess of best approximation. Was established only in normed linear spaces and not in linear metric spaces. This motivated the extension of the notion of strict convexity to linear metric spaces. Various forms of strict convexity and its relation with other spaces have been discussed by sastry and naidu ([15]) and ([16]). Some characterizations of strictly convex linear metric spaces were given by Naran in [9] and [10].

Approximation theory is an old and rich branch of Analysis. The theory is as old as Mathematics itself. The ancient Greeks approximated the area of a closed curve by the area of a polygon. Since the particular examples of approximation often arise from problems of science and technology, they provide proper motivation for the subject of approximation theory. The starting point of approximation theory is the concept of best approximation. Starting in 1853 , P.L. Chebyshev made significant contributions in the theory of best approximation. The problem of best approximation amounts to the problem of finding for a given point x and a given set G of a space X, a point $g_0 \in G$ which is nearest to x amongst all the points of the set G. Such an element g_0 , if it exists , is called a best approximation(or a nearest point or a closest point)to x in G.

1

For most of the available literature in the theory of best approximation , the underlying spaces are normed linear spaces (see e.g. [15],[19],[21],[33],[34],[61],[63],[83] and [94]).In more general spaces, results obtained do not constitute a unified theory as in the case of normed linear spaces. The construction of such a theory upto the present is an open problem although some attempts in this direction have been made by G.C.Ahuja , G.Albinus, E.W.Cheney, N.V.Efimov,T.D.Narang ,Geetha S.Rao, Ivan Singer , S.P.Singh ,S.B.Steckin, Swaran Trehan and many others (see e.g. [3],[4],[5],[19],[84],[85] and the references cited therein). One of our aim I this book is also to make an attempt in this direction.

Fixed point theory plays an important role in functional analysis, differential equations, integeral equations, boundary value problems, statistics,engineering, economics etc. The problem of solving the equation f(x)=0 is equivalent to finding a fixed point of the mapping $y \rightarrow y\text{-}f(y)$. Since finding an exact solution of the equation f(x)=0 is not always possible, approximation theory comes to our rescue and we try to find an approximate solution (which is best possible subject to given constraints).Thus the two subjects of approximation theory and fixed point theory are closely related.

Applications of fixed point theorems to approximation theory are well known. Many results in approximation theory using fixed points are available in normed linear spaces.(see e.g. [8],[9].[21],[53],[86],[87]). Another aim in this book is to investigate applications of fixed point theorems to approximation theory when the underlying spaces are spaces more general than normed linear spaces.

The whole book is divided into eight chapters. In chapter I , we set up some notations, give few definitions and terminology to be used in the subsequent chapters of this book. The chapter II , "strictly convex linear metric spaces and their generalizations" has been divided into two sections . The first section , "some basic properties of strictly convex linear metric spaces", deals with the definition, some examples and properties of strictly convex linear metric spaces . In the second section, "Some Special Linear metric spaces" we discuss linear metric spaces with properties A, B, C, S.C. , P.S.C., B.C., P and P_1 .

Chapter III, "Characterizations of Strictly convex Linear Metric Spaces" deals with some characterizations of strictly convex linear metric spaces. Theorem 3.1

gives necessary and sufficient condition for the real line to be strictly convex. Theorem 3.2 which characterizes strictly convex linear metric spaces in terms of best approximation shows that converse of unicity theorem "Every Convex proximinal set in a strictly Convex linear metric space is Chebyshev" is also true. Theorem 3.3 tells that a linear metric space is strictly convex if and only if all convex subsets of it are semi-chebyshev. Theorem 3.4 shows that a linear metric space is strictly convex if and only if every non-empty locally compact closed convex subset of it is a chebyshev set.

Chapter IV, deals with some results "On Best approximation and Metric Projections". This chapter has been divided into two sections. First section deals with best approximation in pseudo strictly convex metric linear spaces, a notion introduced and discussed by K.P.R. Sastry and S.V.R. Naidu in [69] and [71]. It was shown by Paul C. Kainen et al [38] that the existence of a continous best approximation in a strictly convex normed linear space X and taking values in a suitable subset M of X implies that M has the unique best approximation property. In the first section of this chapter, we extend this result of Paul C. Kainen at al to pseudo strictly convex metric linear spaces.

The second section of this chapter deals with the study of multivalued metric projections in convex metric linear spaces and convex metric spaces. S.B. Steckin [89] proved that if $U_M = \{x \epsilon X; \text{ Card } P_M(x) \leq 1\}$ then $U_M = X$ for every subset M of X iff X is a strictly convex normed linear metric space. This result was extended to strictly convex metric spaces by T.D. Narang [49]. A question that arises is what happens in spaces which are not strictly convex? To answer this, we have discussed in this section, a characterization of multi-valued metric projection P_M in spaces which are not strictly convex. For normed linear spaces which are not strictly convex. This result was proved by loan Serb in [72]. We have also proved in the second section that for non-void proper subset M of a complete convex metric linear space X, P_M cannot be a countable multivalued metric projection. We have also given a characterizaton of the semi-metric linear spaces in terms of finitely-valued metric projections in this section. In [73], it was proved that if M is a strongly proximinal subset of a Banach Space X, then Card $P_M(x) \geq c$ for every xex\M, and the completeness of the space is essential for the validity of the result. In [74], the same result was proved for complete metrizable locally convex spaces i.e. in Frechet

3

spaces. In this section, we have proved that for a strongly proximinal set M in a complete convex metric space (X,d), Card $P_M(x) \geq c$ for all x e X\M.

Chapter V deals with results "On $\mathcal{E}$-Birkhoff Orthogonality and ε-Near Best Approximation". The nation of Birkhoff Orthogonalty, introduced in normed linear spaces in [11], was used to prove some results on best approximation (see [84], p.91). This notion of Orthgonality was extended to metric linear spaces by T.D. Narang and some results on best approximation were proved in [47]. A generalization of Birkhoff Orthogonality [11], called $\mathcal{E}$-Birkhoff Orthogonality, was introduced by Sever Silvestru Dragomir [23] in normed linear spaces and this notion was used to prove a decomposition theorem ([23]-Theorem 3). We have extended this notion of ε-Birkhoff Orthogonality and proved the decomposition theorem in metric linear spaces in section 1.

It was shown by Paul C. Kainen et al [38] that the existence of a continuous ε-near best approximation in a strictly convex normed linear space X and taking values in a suitable subset M implies that M has the unique best approximation property. By extending this result of Paul C. Kainen to convex metric spaces, we have proved in section 2 that for a boundedly compact, closed subset M of a convex metric linear space (X,d) which is also pseudo strictly convex, if for each $\mathcal{E} > 0$, there exists a continuous e-near best approximation $\phi : X \rightarrow M$ of X by M then M is a Chebyshev set. We have extended some other results on ε-near best approximation proved in [38] to metric linear spaces in this section.

Chapter VI deals with results "On $\mathcal{E}$-Simultaneous Approximation and Best Simultaneous Co-approximation". The problem of best simultaneous approximation (b.s.a.) is concerned with approximating simultaneously elements x_1, x_2 of a metric space (X,d) by the elements of a subset G of X. More generally, if a set of elements B is given in X, one might like to approximate all the elements of B simultaneously by a single element of A. This type of problem arises when a function being approximated is not known precisely, but is known to belong to a set. C.B. Dunham [24] seems to be the first to have studied this problem of b.s.a. in normed linear spaces. The study was followed by J.B. Diaz and H.W. McLaughlin, W.H.Ling, Goel at al and many others (see e.g. [2] ,[27], [52], [56], [57] and [66]). R.C. Buck [18] studied the problem of $\mathcal{E}$-approximation which reduces to the problem

4

of best approximation for the particular case when $\varepsilon = 0$. In the first section of this chapter, we have discussed ϵ-simultaneous approximation. Defining ε-simultaneous approximation map $P_{G(\varepsilon)} : X \times X \to 2^G$ ($\equiv$ the collection of all bounded subsets of G) by $P_{G(\varepsilon)}(x_1, x_2) = \{g_0 \in G : d(x_1, g_0) + d(x_2, g_0) \leq r + \varepsilon\}$ where r $=$ $\inf\{d(x_1, g) + d(x_2, g) : g \in G\}$ and $P_{G(\varepsilon)}(F) =$ $\{g_0 \in G : \sup_{y \in F} \leq \inf_{g \in G} \sup_{y \in F} d(y, g) + \varepsilon\}$, We have also proved the upper semi-continuity of the maps $P_{G(\varepsilon)}(x_1, x_2)$ and $P_{G(\varepsilon)}(F)$ and the convexity, boundedness, closedness and the starshpaedness of the sets $P_{G(\varepsilon)}(x_1, x_2)$ and $P_{G(\varepsilon)}(F)$ in this section.

The second section of this chapter deals with Best Simultaneous Co-approximation. This concept of best simultaneous co-approximation was introduced and discussed in normed linear spaces by C. Franchetti and M. Furi [26] in 1972. The study was taken up later by T.D. Narnag, P.L. Papini, Geetha S. Rao, Ivan Singer and few others (see [56],[57]. Generalizing the concept of best co-approxiamtion, Geetha S. Rao and R. Sarvanan studied the problem of best simultaneous co-approximation in nomed linear spaces in [64]. In the second section, we have studied the problem of best simultaneous co-approximation in convex metric linear spaces and convex metric spaces, thereby extending some of the results proved in [64]. We have also given some properties of the set $S_G(x, y)$ i.e. the set of all best simultaneous co-approxiamtions to x,y in G. We have proved that for a convex metric space (X,d) G a convex subset of X and x, y $\in$ X, the set $S_G(x, y)$ is a convex set. We have also proved the upper semi-continuity of the mapping $S_G : \{(x,y): x, y \in X\} \to 2^G$ in totally complete metric linear spaces (a notion introduced by T.D. Narnag [50]).

Chapter VII deals with "Fixed points and Approximation". We have divided this chapter into three sections. First section deals with fixed points in pseudo strictly convex metric linear spaces. The problem of fixed points of non-expansive mappings have been extensively discussed in strictly convex normed linear spaces (see e.g. [36]). It is known (see e.g. [12] Theorem 6, p.243) that for a closed convex subset K of a strictly convex normed linear space X and a non-expansive mapping T:K $\to$ X, the fixed point set (possibly empty) of T is a closed convex set. We have extended this result to pseudo strictly convex metric linear spaces. Using fixed point theory,

Brosowski [14] and Meinardus [43] established some interesting results on invariant approximation in normed linear spaces. Later various researchers obtained generalizations of their results (see e.g. [36] and the references cited therein). The object of the second section of this chapter is to extend and generalize some results of Brosowski [14], Hicks and Humphries [31], Khan and Khan [39], and Singh [86], [87] in metric spaces having convex structure and in metric linear spaces having convex structure and in metric linear spaces having strictly monotone metric. Considering a subset C of a metric linear space with strictly monotone metric d and a non-expansive mapping T on $P_C(x) \cup \{x\}$ where x is a T-invariant point, we have proved the existence of an x_0 in the set $P_C(x)$ satisfying certain conditions. In this section we have also established a result on invariant approximation in strictly convex metric spaces. In the third section we have given an application of a fixed point theorem to ε-simultaneous approximation in convex metric spaces.

Chapter VIII deals with "Non-expansive Retracts in Convex Metric Spaces". To generalize a theorem of Belluce and Kirk [10] on the existence of a common fixed point of a finite family of commuting non-expansive mappings, Ronald E.Bruck Jr. [17] studied some properties of fixed-point sets of non-expansive mappings in Banach spaces. In this chapter, we extend some of the results of [17] to convex metric spaces. We also prove that the fixed point set of a non-expansive mapping satisfying conditional fixed property (CFP) is a non-expansive retract of C and hence metrically convex.

The book ends with a list of references

--- X ---

CHAPTER-I
PRELIMANARIES

In this chapter, we set up some notations and terminology to be used in the subsequent chapters. Throughout this book, the underlying field will be either the field of real numbers or the field of complex numbers. R will denote the set of real numbers; R^+ will denote the set of non-negative real numbers; C, the set of complex numbers; iff for if and only if ; B(x,r) , a closed sphere with centre x and radius r ; ∂C , the boundary of a set C; X\E , the set of those points of X which are not in E; I ,the closed unit interval [0,1]; card A, the number of infinite countable set and c, the cardinality of the interval. The symbol $\in$ will stand for belongs to, iff for if and only if, s.c. for strictly convex, n.1.s. for normed linear space, dim A for dimenstion of a set A, f.d. for finite dimensional, inf. for infimum, sup. for supremum, min. for minimum, max. for maximum, int. for interior, R^n for the n-dimensional Euclidean space, C^n for n-dimensional unitary space. $\emptyset$ for the empty set, conv(A) for the convex hull of a set A, Cl A for the closure of a set A, ∂A denotes the topological boundary of A, E/G for the quotient space of E by G, E\G for the complement of a set G in E, [x] for the subspace generated by an element x, R[f (x)] for the real part of f(x), Im[f(x)] for the imaginary part of f(x), d(x, G) for the distance of x from a set G, line segment [x, y] for the set $\{\alpha x + (1-\alpha)y, 0 \le \alpha \le 1\}$,] x, y [for the open line segment $\{\alpha x + (1-\alpha)y, 0 < \alpha < 1\}$, $x_n \rightharpoonup x$ for x_n converges to x weakly. In a linear metric space (X, d), B [0, r] = {x $\in$ X, d (0, x) $\le$ r} will stand for the closed ball with centre 0 and radius r, B (0, r) = {x $\in$ X, d (0, x) < r} will stand for the open ball with centre 0 and radius r and the functional d (0, ·) = |·| defined on X is called quasinorm of X . The numbers within square barackets indicate references cited at the end of the dissertation. Other notations will be given whenever these occur .

Now we give a few definitions which will frequently occur in the rest of the chapters of this book.

Definition 1.1 A subset A or a linear space X is said to be **convex** if with any two points x, y of A , it contains the line segment joining two points i.e. x , y ∈ A , $\alpha \in]0,1[$ imply αx + (1- α)y ∈ A. It is said to be **mid-point convex** if $\frac{1}{2}x + \frac{1}{2}y \in A$ y ∈ A for any two points x, y ∈ A.

Definition 1.2 If A is any set in a linear space X then intersection of all the convex sets containing A is called the **convex hull** of A.

Definition 1.3 A set V in a linear space X is said to be **linear manifold** if it is of form $V = x_0 + G = \{ x_0 + g : g \in G \}$, where $x_0 \in X$ and G is a linear subspace of X i.e. a translate of a linear subspace of X is called a **linear manifold.**

A closed linear manifold H⊂X is called a **hyperplane** if there exists no closed linear manifold $H_1 \subset X$ such that $H \subset H_1$ and H≠ X i.e. H is the maximal closed linear manifold in X.

Definition 1.4 A set of the form {x ∈ L, r (x)≥ ∝} where L is an n-dimensional subspace of a linear space X (n, a natural number, $1 \leq n \leq \dim X \leq \infty$) f is a linear functional on L, and α is a real number, is called an (closed) **n-dimensional half plane.**

Definition 1.5 A subset A of a linear space X is called **symmetric** if {-x : x ∈ A} = A.

Definition 1.6 A set X with a family β of subsets of X is called a **topological space** if β satisfies the following conditions:

   (a)   The empty set ∅ and the whole set X belong to β.

   (b)   The union of any number or members of β is again a member of β.

   (c)   The intersection of any finite number of members of β is again a member of β.

The family β is called a **topology** for X, and the members of β are called open sets of X in this topology.

Definition 1.7 A vector space X over a field K, together with a Housdorff topology β is called a **topological linear space** if the vector space operations (x, y) → x + y from X x X into X and (α , x) → α x from K x X into X are continous.

8

Definition 1.8 A topological linear space is said to be **locally convex** if it has a base of convex neighbourhoods.

Definition 1.9 A linear space X is called a **linear metric space** if it is a topological linear space with topology derived from an invariant metric i.e. $d(x_1 + y, x_2 + y) = d(x_1, x_2)$ for every choice of x_1, x_2 and y in X.

        Equivalently, a metric space (X, d) is said to be a linear metric space if

(i)   It is a linear space.

(ii)   Addition and scalar multiplications are continuous i.e. for

        $<x_n> \rightarrow x, <y_n> \rightarrow y, <\alpha_n> \rightarrow \alpha,$

        $<x_n + y_n> \rightarrow x + y$ and $\alpha_n x_n> \rightarrow \alpha x$, and

(iii)   d is translation invariant i.e. $d(x_1 + y, x_2 + y) = d(x_1, x_2)$ for every choice of x_1, x_2, y, in X.

Definition 1.10 A linear metric space (X, d) is said to be **bounded liner metric space** if the metric d is bounded i.e. there exists r > 0 such that

$$r = \sup_{x \in X} d(x, 0)$$

Definition 1.11 A linear metric space (X, d) is said to be **strongly locally convex** if each open sphere in it is a convex set.

Definition 1.12 A linear space X is said to be **normed linear space** if to each x ∈ X, there is assigned a unique real number, which we denote by $\|x\|$, satisfying the following properties:

(i) $\|x\| \geq 0$ and $\|x\| = 0$ iff x = 0

(ii) $\|x + y\| \leq \|x\| + \|y\|$

(iii) $\|\alpha x\| = \|\alpha\| + \|x\|$ for all x, y ∈ X and for all scalars α.

Definition 1.13 A normed linear space $(X, \|\cdot\|)$ is said to be **strictly convex** if for any two points x and y or X with $\|x\| = \|y\| = 1$, $\left\|\dfrac{x+y}{2}\right\| < 1$ unless x = y.

Definition 1.14 The **conjugate space** or **dual space** of a n.l.s. X, denoted by X^*, is the space of all continuous linear functionals on X with the usual linear space operations and the norm defined by $\|f\| = \sup_{\substack{\|x\| \leq 1 \\ x \in X}} |f(x)|$

9

Definition 1.15 A normed linear space X is said to be **reflexive** if $X^{**} = X$, where X^{**} stands for the **second conjugate space** of X.

Definition 1.16 A n. l. s. X is said to be **smooth** if every element x of X with $\|x\| = 1$ has a unique support hyperplane to the open unit ball $B(0, 1) = \{ x : x \in E, \|x\| < 1\}$.

Definition 1.17 A normed linear space X is said to be **strictly normed** if the relation $\|x + y\| = \|x\| + \|y\|$, x, y $\in$ X\\{0} implies the existence of a number C > 0 such that y = C x Such a norm is called a **strict norm.**

Definition 1.18 An element x of a n.l.s. X is said to be **orthogonal** to an element y of X, x⊥y, if $\|x+\alpha y\| \geq \|x\|$ for every scalar α. x is said to be **orthogonal to a subset** G of X, x⊥G, if x⊥y for all y $\in$ G.

Definition 1.19 Let X be a linear space. A mapping

$$<\star, \star> \quad : X \times X \longrightarrow K, \quad \text{the}$$

the field or scalars is said to be an **inner product** on X if the following properties hold for all x, y, z $\in$ X and for all scalars α, β $\in$ K

(i) $< x, x > \geq 0$ for all $x \in K$

(ii) $< x, y > = < \overline{y, x} >$ where $< \overline{y, x} >$ denotes the complex conjugate of $< y, x >$

(iii) $< \alpha x + \beta y, z > = \alpha <x, z > + \beta <y, z>$

The pair $(X, <\cdot, \cdot>)$ is called an **inner product space.**

Definition 1.20 A complete inner product space is called a **Hilbert space.**

Definition 1.21 Let $(X, \Im)$ and (Y, β) be two topological spaces. Then a mapping $f : X \to Y$ is called a **homeomorphism** if

(i) f is one-one

(ii) f is onto

(iii) f is continous

(iv) f^{-1} is continous

Definition 1.22 If E and F are two normed liner spaces over the same field K then a mapping $T : E \quad > F$ is called a **linear transformation** if it satisfies the following properties:

(i) $T (x + y) = Tx + Ty$

(ii) $T (\alpha x) = \alpha T (x)$ for any x, y ϵ E and for all $\alpha \epsilon$ K.

Linear transformation T is said to be bounded if

$$\|T\| = \sup_{\substack{\|x\| \leq 1 \\ x \in E}} \|T(x)\| \quad \text{is finite}$$

Definition 1.23 A subset A of a metric space (X, d) is said to be **metrically bounded** or **d-bounded** if sup {d (x, y), x, yϵ A } is finite.

Definition 1.24 A subset G of a metric space (X,d) is said to be **proximinal** if for each $x \in X$ there exists a point g_0 in G which is nearest to x i.e.

$$d (x, g_0) = d(x,G) \equiv \inf \{d(x , g) : g \in G\} \tag{1.1}$$

The term 'proximinal' was proposed by Raymond Killgrove (see Phelps [59], p. 790).

Every element $g_0 \in G$ satisfying (1.1) is called **an element of best approximation** of x by the elements of G or a **nearest point** or a **closest point** to x in G.

We shall denote by $P_G(x)$, the set of all best approximations to x in G i.e.

$$P_G(x) = \{ g_0 \in G : d (x, g_0) = d(x, G)\}.$$

Thus G is **proximinal** if $P_G(x)$ is non-empty for each $x \in X$.

Since

$$P_G(x) = \begin{cases} x, & x \in G \\ \emptyset, & x \in \overline{G} / G, \end{cases} \tag{1.2}$$

it follows that every proximinal set is closed.

The following example shows that a closed set need not be proximinal.

Example 1.1 Let $C_o = \{<a_n> : a_n \in F \text{ (F=R or C)}, a_n \to 0\}$ with

$$d(<a_n>,<b_n>) = \sup_n d(a_n, b_n)$$

Let $M = \{<a_n> \in C_0 : \sum_{n \in N} 2^{-n} a_n = 0\}$

11

then M is a closed infinite dimensional subset of C_o and if $x=<b_n>\notin M$, then there is no $m\notin M$ such that
$$d(x,m) = d(x,M).$$

In view of (1.2), in order to exclude the trivial case when elements of best approximation do not exist, throughout while discussing $P_G(x)$, we shall assume, without special mention that
$$\overline{G} \neq X.$$

In case $P_G(x)$ is exactly singleton (atmost singleton) for each $x \in X$, we have the following:

Definition 1.25 A set G in a metric space (X, d) is said to be **Chebyshev** or **uniquely proximinal (semi-Chebyshev)** If $P_G(x)$ consists of exactly one (atmost one) point for each x in X i.e. for each $x \in X$ there exists exactly one (atmost one) $g_o \in G$ such that $d(x, g_0)=d(x,G)$.

Example 1.2 [46]. Let $X = R^2$ with usual metric and
$$G = \{x,y\} : x = -\sqrt{1-y^2}, -1 \le t \le 1\}$$

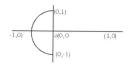

If x=(1,0) then
$$P_G(x)= \{(0,1),(0,-1)\}$$

If x=(0,0) then $P_G(x)=G$.

The set G is proximinal but not Chebyshev.

Example 1.3 [46] A closed bounded interval [a,b] on the real line is a Chebyshev set.

Definition1.26 The mapping which takes each point x of the space X to those points of the set G which are nearest to x is called **a best approximation map or nearest point map** or **a metric projection**.

Definition1.27 A set G in a metric space (X , d) is said to be **approximatively compact** (Effimov and Steckin [25]) if for every $x \in X$ and every sequence $< g_n >$ in G with

$$\lim_{n \to \infty} d(x, g_n) = d(x, G) \tag{1.3}$$

there exists a sequence $< g_{n_i} >$ converging to an element of G. Any sequence $< g_n >$ satisfy (1.3) is called a **minimizing sequence** for x in G.

An approximatively compact set in a metric space is proximinal (Effimov and Steckin [25]) and since equation (1.2) implies that every proximinal set is closed ,it follows that every proximinal compact set is closed, it follows that every approximatively compact set is closed. But a proximinal set need not be approximatively compact (Singer [84], p.389).

Definition1.28 A set G in a metric space (X , d) is said to be **boundedly compact** (Klee [50]) if every bounded sequence in G has a subsequence converging to a point of the space X. Equivalently, if the closure of $G \cap B$ is compact for each closed ball B in X.

In a metric space, every boundedly compact, closed set is approximatively compact (Effimov and Steckin [25]) and hence proximinal.

Definition1.29 Let (X,d) be a metric space and G a non-empty subset of X. An element $g_x \in G$ is called a **best co-approximation** to x if

$$d(g_x, g) \le d(x, g) \text{ for every } g \in G.$$

The set of all best co-approximation to $x \in X$ is denoted by $R_G(x)$.

13

Definition1.30 Let (X,d) be a metric space and G a non-empty subset of X. An element $g_0 \in G$ **is an element of best simultaneous approximation (b.s.a.) to** $x_1, x_2 \in X$ from G if

$$d(x_1, g_0) + d(x_2, g_0) = \inf\{d(x_1, g) + d(x_2, g) : g \in G\}$$

The set of all best simultaneous approximations to $x_1, x_2 \in X$ from G is denoted by $P_G(x_1, x_2)$.

Definition1.31 Let (X , d) be a metric space , G a non-empty subset of X and F a non-empty bounded subset of X. . An element $g_0 \in G$ is called **an element of best simultaneous approximation of F with respect to G** if

$$\sup_{y \in F} d(y, g_0) \le \inf_{g \in G} \sup_{y \in F} d(y, g)$$

The set of all best simultaneous approximation to F with respect to G is denoted by $P_G(F)$.

Definition1.32 [11] An element x of a normed linear space X is said to be **orthogonal** to $y \in X$, $x \perp y$ if

$$d(x, 0) \le d(x, \alpha y)$$

For every scalar α.

Correspondingly, we say that an element x of a metric linear space (X,d) is orthogonal to a subset M of X, if $x \perp y$ for each y in M.

Definition1.33 Let (X,d) be a metric space and C a subset of X. A mapping T:C→ X is said to be **non-expansive** if $d(Tx, Ty) \le d(x, y)$ for all x,y $\in$ C. The set F(T) = $\{x \in X : T(x) = x\}$ is called the **fixed point set** of the mapping T and a point of F(T) is called a **T-invariant point** in X.

Definition1.34 For two non-empty sets A and B, a mapping

T:A→B is called a retraction of A onto B if

14

(a)   B is a subset of A,

(b)   Tx = x for all x ∈ B.

The set B is said to be a retract (non-expansive retract) of A if there exists a retraction (n0on-expansive retraction) of A onto B.

Remark 1.1 The metric projection $\pi_G : X \rightarrow G$ defined by

$\pi_G(x) = \{g \in G : d(x,g) = d(x,G)\}$ is a retraction of X onto G.

Definition1.35 For a metric space (X,d), a continuous mapping $W : X \times X \times I \rightarrow X$ is said to be **convex structure** on X if for all x, y $\in X \lambda, \; l \in$,

$$d\, u\, (W, x\, (y\, \lambda, \; \ni)\lambda d\, u\, (x\, ,+) \; -(\lambda\, d\,)u\, (y,$$

for all $u \in X$. The metric space (X, d) together with a **convex structure** is called a **convex metric space** [90].

Clearly, a normed linear space or any convex subset of it is a convex metric space with $W(x, y, \lambda) = \lambda x + (1-\lambda)y$. But a linear metric space is not necessarily a convex metric space. There are many convex metric spaces (see Takashashi [90]) which cannot be embedded in any normed linear spaces. We give two preliminary examples here.

Example 1.4 [90]. Let I be the unit interval [0,1] and X be the family of closed intervals $[a_i, b_j]$ such that $0 \le a_i \le b_j \le 1$. For $I_i = [a_i, b_i]$, $I_j = [a_j, b_j]$ and $\lambda (0 \le \lambda \le 1)$, we define a mapping W by $W(I_i, I_j, \lambda) = [\lambda a_i + (1-\lambda)a_j, \lambda b_i + (1-\lambda)b_j]$ and define a metric d in X by the Hausdorff distance i.e.

$$d(I_i, I_j) = \sup_{a \in I} \left\{ \inf_{b \in I_i}(|a-b|), \inf_{c \in I_j}|a-c| \right\}$$

Example 1.5 [90]. The linear space L which is also a metric space with the following properties:

(1)   x, y ∈ L, d (x, y) = d (x-y, 0);

(2)   For x, y ∈ L and λ $(0 \le \lambda \le 1)$,

$$d(\lambda x + (1-\lambda)y, 0) \le \lambda d(x, 0) + (1-\lambda)d(y, 0).$$

Definition1.36 A convex metric space (X,d) is said to be **strictly convex** [49] if for every x, y $\in$X and r>0,d(x,p) $\leq$r, d(y,p) $\leq$r imply d(W(x, y, λ),p)<r unless x=y , where p is arbitrary but fixed point of X and $\lambda \in$I.

Definition1.37 A non-empty subset C of a convex metric space (X,d) is said to be

(a)   **Starshaped** [90] if there exists a p $\in$ C such that W(x ,p, λ) $\in$ C for every $\lambda \in$I and for every x$\in$ C . Such a p is called a **starcentre** of C.

(b)   **Convex** [90] , if W(x, y, λ) $\in$C whenever x,y $\in$ C and $\lambda \in$I.

Clearly , a convex set is starshaped with respect to each of its points.

Definition1.38 A convex metric space (X,d) is said to satisfy **property (I)** [8], if for all x, y $\in$X and $\lambda \in [0,1]$

$$d(W(x,p,\lambda),W(y,p,\lambda)) \leq \lambda d(x,y)$$

A convex metric space (X,d) is said to satisfy **property (I*)** if for all x, y $\in$ X and $\lambda \in [0,1]$

$$d(W(x,p,\lambda),W(y,p,\lambda)) \leq (1-\lambda)d(x,y)$$

Clearly properties (I) and (I*) hold in normed linear spaces and in linear metric spaces satisfying

$$d(\lambda x + (1-\lambda)y,0) \leq \lambda d(x,0) + (1-\lambda)d(y,0)$$

Definition1.39 A normed linear space (X, $\|\cdot\|$) is said to be **strictly convex** if for any two points x and y of X and r> 0 with $\|x\| \leq r$, $\|y\| \leq r$ imply $\|(x+y)/2\| <$ r unless x =y.

Definition1.40 A normed linear space (X, $\|\cdot\|$) is said to be **pseudo strictly convex (P.S.C.)** if given x $\neq$0 , y $\neq$0,$\|x+y\| = \|x\| + \|y\|$ implies y=tx for some t > 0.

For normed linear spaces , strict convexity and pseudo strictly convexity are equivalent (see e.g. [12] p. 122, [33] , [34] and [71]). Some authors use for such spaces the term strictly normed space or rotund space (see e.g. [20]).

Geometrically , strict convexity means that the spheres of the space contain no line segment on their surfaces. In such a space, if the sum of the lengths of two sides

16

of a triangle is equal to the length of the third side, the triangle is degenerate. Three – dimensional strictly convex space is the one having a "football" shaped unit ball.

A very good account of strict convexity can be found in [37].

The notion of strict convexity was extended to metric linear spaces in [1] as under:-

Definition 1.41 A metric linear space (X, d) is said to the strictly convex if

$d(x,0) \leq r$, $d(y,0) \leq r$ imply $d((x+y)/2,0) < r$ unless $x=y$; $x, y \in X$ and r is any positive real number.

--- X ---

17

CHAPTER-II
SRICTLY CONVEX LINER METRIC SPACES AND THEIR GENERALIZATION

The concept of strict convexity was extended to linear metric spaces by G.C.Ahuja, T.D. Narang and Swaran Trehan in [1]. This chapter "Strictly Convex Linear Metric Spaces and their Generalizations" has been divided into two sections. The first section deals with the definition and some examples of strictly convex linear metric spaces. In this section, we also prove some properties of linear metric spaces. In the second section, we give specail linear metric spaces i.e. linear metric spaces with properties : A, B, C, S.C., P.S.C., B.C., P and P_1 .

2.1 SOME BASIC PROPERTIES OF STRICTLY CONVEX LINEAR METRIC SPACES

We begin with the notion of strictly convex linear metric space as introduced by Ahuja, Narang and Trehan in [1].

Definition 2.1.1 [1] A linear metric space (X, d) is said to be **strictly convex** if $d(x,0) \le r, d(y,0) \le r$ imply $d(\frac{x+y}{2}, 0) < r$ unless $x = y$, $x, y \in X$ and r is any positive real number.

Next we give an example of a strictly convex linear metric space.

Example 2.1.1 [1] The set R of real numbers with metric d defined by $d(x, y) = \frac{|x-y|}{1+|x-y|}$ is a strictly convex linear metric space. This can be seen as follows:

Lex x and y be two distinct points of R with $d(x, 0) \le r$, $d(y, 0) \le r$. These give

$$|x| \le \frac{r}{1-r}, \quad |y| \le \frac{r}{1-r}$$

Strict convexity of $|\cdot|$ implies $\left|\dfrac{x+y}{2}\right| < \dfrac{r}{1-r}$, which in turn implies

$d\left(\dfrac{x+y}{2}, 0\right) < r$

Remark 2.1.1 [1] In fact we can say a little more viz. if $(X, \|\cdot\|)$ is a strictly convex normed linear space then the linear metric space (X, d) where d is defined as

$$d(x,y) = \dfrac{\|x-y\|}{1+\|x-y\|}$$

is strictly convex.

The following example shows that even a finite dimensional liner metric space need not be strictly convex.

Example 2.1.1 [1] Consider (R^2, d) where d is defined as

$d(x,y) = \max\{|x_1-y_1|, |x_2-y_2|\}$

$x = (x_1, x_2), \ y = (y_1, y_2).$

Let $x = (1, 1)$ and $y = (1, 0)$ then $d(x, 0) = 1$, $d(y, 0) = 1$ and also $d\left(\dfrac{x+y}{2}, 0\right) = 1$

. Therefore (R^2, d) is not strictly convex.

The following theorem tells that certain type of linear metric spaces can never be strictly convex.

Theorem 2.1.1 [67] A non-zero bounded linear metric space in which the metric attains its superemum is not strictly convex.

Proof : Let (X, d) be a non-zero bounded linear metric space such that d attains its superemum, say r. Then $r > 0$ and there exists $z \in X$ such that $d(z, 0) = r$. Take $x = \dfrac{z}{2}$ and $y = \dfrac{3}{2}z$. Then $d(x, 0) \le r$, $d(y, 0) \le r$ and $x \ne y$, but

$d\left(\dfrac{x+y}{2}, 0\right) = d(z, 0) = r$. Hence (X, d) is not strictly convex.

It is well known (see [60]) that Every Convex Proximinal set in a Strictly Convex Normed Linear Space is Chebyshev. We show below that a similar result holds in a strictly convex linear metric space.

19

Theorem 2.1.2 [1] A convex proximinal set in a strictly convex linear metric space is Chebyshev.

Proof: Let G be a convex proximinal set in a strictly convex linear metric space (X, d) and let p be any arbitrary point on X. Since G is proximinal, there exists $g_1^* \in G$ such that $d(p, g_1^*) = d(p, G) \equiv r$ (say).

Let, if possible, there exists $g_2^* \in G$ such that $d(p, g_2^*) = r$. Invariance of the metric d implies that

$$d(p-g_1^*, 0) = d(p - g_2^*, 0) = r.$$

Since X is strictly convex, we have

$$d\left(\frac{(p - g_1^*) + (p - g_2^*)}{2}, 0\right) < r \text{ unless } p - g_1^* = p - g_2^*$$

i.e. $d\left(p, \frac{g_1^* + g_2^*}{2}\right) < r$ unless $g_1^* = g_2^*$. Since $\frac{g_1^* + g_2^*}{2} \in G$, definition of r

implies that $g_1^* = g_2^*$. Hence G is Chebyshev.

Now, we give a lemma to be used in Theorem 2.1.3 which shows that strictly convex linear metric spaces are strongly locally convex – a notion introcduced by T.D. Narang in [48].

Lemma 2.1.1 [67] Let (X, T) be a topological vector space and S be a non-empty closed subset of X such that x, y ϵ ∂S (boundary of S) and x ≠ y imply (x, y) ∩ S=Ø. Then S is convex.

Proof: Suppose S is not convex. Then there exist x, y $\in$ S, x ≠ y such that (x, y) ∩ $S' = $Ø ($S'$ is the compement of S in X).

Let A = { t$\in$ (0, 1) : tx + (1-t) y $\in$ S' } Then A is anon-empty subset of R. Let B be a component of A. Then there exist α, β $\in$ R such that α < β = (α, β). write

$$z_1 = \alpha x + (1-\alpha) y \text{ and } z_2 = \beta x + (1-\beta) y$$

Then, clearly z_1, z_2 are distinct points of ∂S and (z_1, z_2) ∩ S =Ø which contradicts the hypothesis.

Theorem 2.1.3 [67] In a strictly convex linear metric space, the balls are convex.
Proof From lemma 2.1.1, it is clear that closed balls with centre at the origin and hence the open balls with centre at the origin are convex. Since every ball is a translate of a ball with centre at the origin, the result is immediate.

2.2 SOME SPECIAL LINEAR METRIC SPACES

In this section, we shall discuss some special linear metric spaces i.e. linear metric spaces with properties A, B, C, S.C., P.S.C., B.C., P and P_1 and the relationships of A, B, C, P.S.C., B.C., P and P_1 with S.C.

We say that a linear metric space (X, d) has the
Property :-

A : Given $r > 0$, $\varepsilon > 0$ there exists $\delta > 0$ such that
$B[0, r + \delta] \subset B[0, r] + B[0, \varepsilon]$

B : Given $r > 0$, $\varepsilon > 0$ there exists $\delta > 0$ such that
$d(x, 0) > r\text{-}\delta \Rightarrow \sup \{d(x + z, 0) : d(z, 0) < \varepsilon\} > r$.

C: Give $r > 0$, $\varepsilon > 0$ there exists $\delta > 0$ such that $r < d(x, 0) < r + \delta \Rightarrow$ there exists y, z such that $d(y, 0) = r$, $d(z, 0) < \varepsilon$ and $x = y + z$.

S.C. $r > 0$, $x \neq y$, $d(x, 0) \leq r \Rightarrow d(\dfrac{x + y}{2}, 0) < r$.

P.S.C. $x \neq 0$, $y \neq 0$, $d(x + y, 0) = d(x, 0) + d(y, 0) \Rightarrow y = tx$ for some $t > 0$.

P. A linear metric space (X, d) is said to have property (P) if the nearest point mapping shrinks distances whenever it exists.

B.C. $r \geq 0$, $d(x, 0) = d(y, 0) = r \Rightarrow d(\dfrac{x + y}{2}, 0) \leq r$.

P_1 A linear metric space (X, d) is said to have property (P_1) if for every pair of elements $x, z \in X$ such that $d(x + z, 0) \leq d(x, 0)$ there exist constants $b = b(x, z) > 0$, $c = c(x, z) > G$ such that $d(y + c z, 0) \leq d(y, 0)$ for $d(y, x) \leq b$.

Lemma 2.2.1 [69] Let $f : R^+ \to R^+$ be strictly increasing function such that (X, $f \circ d$) is a linear meric space. Then (X, d) has S.C. $\Leftrightarrow$ (X, $f \circ d$) has S.C.
Proof Let $r > 0$ and $(f \circ d)(x, 0) \leq r$, $(f \circ d)(y, 0) \leq r$
i.e. $f[d(x, 0)] \leq r$, $f[d(y, 0)] \leq r$.
We may assume that there exists $z \in X$ such that $f[d(z, 0)] \geq r$. Since

21

f (d(t z, 0)) is a continuous function of t on R and hence for some $t \in \,]\,0, 1\,[\,$, f (d (tz, 0)) = r so that $f^{-1}(r)$ exists.

Clearly $f^{-1}(r) > 0$

$$d\,(x, 0) \le f^{-1}\,(r), d\,(y, 0) \le f^{-1}\,(r)$$

and so by strict convexity of d, $d(\dfrac{x + y}{2}, 0) < f^{-1}\,(r) \Rightarrow$

$f\,[d(\dfrac{x + y}{2}, 0)] < r$ as f is strictly increasing

Since f^{-1} is strictly increasing and $d = f^{-1} \circ (f \circ d)$, the other implication follows from the first.

As above, the following lemma can be easily established.

Lemma 2.2.2 [69] Let $f : R^+ \to R^+$ be a strictly increasing function such that
(i) $f\,(s + t) \le f\,(s) + f\,(t)$ for all s, $t \in R^+$ and
(ii) $(X, f \circ d)$ is a linear metric space. Then (X, d) has P.S.C. $\Rightarrow$ (X, $f \circ d$) has P.S.C.

Proof :   Let $(f \circ d)\,(x + y, 0) = (f \circ d)\,(x, 0) + (f \circ d)\,(y, 0)$

$$
\begin{array}{rl}
 & f \circ d)\,(x + y, 0) \\
= & f\,(d\,(x + y, 0)) \\
\le & f\,(d\,(x, 0) + d\,(y, 0)\,) \\
f\,(d\,(x, 0) + d\,(y, 0)) \quad = & f\,(d\,(x + y, 0)\,)
\end{array}
$$

Now since r is a strictly increasing function, we have
$$d\,(x, 0) + d\,(y, 0) = d\,x + y, 0)$$
which implies that y = tx for some t > 0 and so (X, $f \circ d$) has P.S.C.

Suppose f: $R^+ \to R^+$ is such that (X, $f \circ d$) is a linear metric space. Then f does not satisfy that condition $f\,(s + t) < f\,(s) + f\,(t)$ for all s, $t \in R^+$ as is evident from the following example:

22

Example 2.2.1 [69] The function d defined by d $(x, y) = |x\text{-}y|^{1/2}$ is a linear lmeric on R. If we define $f: R^+ \to R^+$ by $f(t) = t^2$, then $f \circ d$ is the usual metric on R. Clearly f does not satisfy the condition.

$f(s + t) \le f(s) + f(t)$ for all s, $t \in R^+$

Given two linear metrics on a linear space X, their Euclidean combination on X x X is a linear metric while the Euclidean combination of two strictly convex norms on X is strictly convex. On X x X, the same need not be true in the case of linear metrics as the following example shows.

Example 2.2.2. [69] Consider the strictly convex linear metric space (R, d_1), where $d_1(s, t) = |s\text{-}t|^{1/2}$ for all s, $t \in R$. Then

$$d((x_1, y_1), (x_2, y_2)) = [|x_1 - x_2| + |y_1 - y_2|]^{1/2}$$

is the Euclidean combination of d_1 with itself and is a linear metric on R^2. Clearly d $((1, 0), (0, 0)) = d((0, 1), (0, 0)) = 1$

and d $((\frac{1}{2}, \frac{1}{2}), (0, 0)) = 1$.

Hence (R^2, d) is not strictly convex.

Each of the following two examples shows that if (X, d) has P.S.C. then it need not have S.C. In the first example the balls are convex whereas in the second example all the balls are not convex.

Example 2.2.3 [69] Let $f: R^+ \to R^+$ be defined by

$$f(t) = \begin{cases} t & if & 0 \le t \le 1 \\ 1 & if & t > 1 \end{cases}$$

and d be the linear metric on R defined by d $(0, t) = f(|t|)$ for all $t \in R$. Then (R, d) has (B.C.) and P.S.C. but not S.C.

Example 2.2.4 [69] Let $f: R^+ \to R^+$ defined by

$$f(t) = \begin{cases} t & if & 0 \le t \le 1 \\ \frac{1}{2}(1+\frac{1}{t}) & if & t \ge 1 \end{cases}$$

and d be the linear metric on R defined by d $(0, t) = f(|t|)$ for all

$t \in R$. Then (R, d) has P.S.C. but neither (B.C.) nor (A) nor (B).

Now we give two more examples giving the relation between S.C., (A) and (B).

Example 2.2.5 [67] Define f : $R^+ \to R^+$ defined by

$$f(t) = \begin{cases} t, & \text{if } 0 \le t \le 1 \\ 1, & \text{if } 1 < t < 2 \\ \dfrac{t}{2}, & \text{if } t \ge 2 \end{cases}$$

and   d: R x R → R^+ by

d (x, y) = f (|x- y|).

Then (R, d) is a totally complete Linear metric space such that all of its balls are convex but it is not strictly convex. Further it satisfies (A) but not (B) even though d is unbounded.

Example 2.2.6 [67] Define f : $R^+ \to R^+$ by

$$f(t) = \frac{t}{1+t}$$

and   d: R x R → R^+

by   d(x, y) = f (|x − y|).

Then (R, d) is a bounded strictly convex linear metric space satisfying (A) but not (B).

Next, we show that a totally complete linear metric space satisfies (A) and totally complete linear metric space, in presence of S.C., satisfies (B). This is the essence of our next theorem.

Theorem 2.2.1 [67] The following hold:
   (i)   A totally complete linear metric space satisfies (A).
   (ii)   A totally complete strictly convex linear metric space satisfies (B).
Proof (i) Let (X, d) be a totally complete linear metric space. Let r> 0 and $\varepsilon > 0$ suppose there does not exist $\delta > 0$
such that

$$B[0, r + \delta] \subset B[0, r] + B[0, \varepsilon].$$

Then there exists $z_n \in X$ such that

24

$$d(z_n, \ 0) > r + \frac{1}{n} \text{ and}$$

$$z_n \notin B[0, r] + B\,[0, \ \varepsilon]$$

so that

$$d(z_n \ , \ 0) > r$$

Since (X, d) is totally complete, there exists a convergent subsequence $\{Z_{n_i}\}$ of $\{z_n\}$ with limit, say, z. Then d (z, 0) = r and hence z is an interior point of

$$B[0, r] + B[0, \varepsilon]$$

so that it contains infinitely many z_n which is a contradiction.

(ii) Let (X, d) be a totally complete, strictly convex linear metric space. Let r > 0 and ε >0.

Suppose there does not exist $\delta > 0$ such that d (x, 0) > r - δ implies

$$\sup \ \{d \ (x + z, 0) \ , d \ z, 0) < \varepsilon \ \} > r.$$

Then there exists $x_n \in$ X Such that

$$d(x_n, 0) > r - \frac{1}{n}$$

and

$$\sup \ \{d \ (x_n + z, 0) : d(z, 0) < \varepsilon\} \le r.$$

Since (X, d) is totally complete, there exists a convergent subsequence $\{x_{n_i}\}$ of $\{x_n\}$ with limit, say, x. Then d(x, 0) =r and

$$\sup\{d(x + z, 0) : d \ (z, 0) < \varepsilon\} \ \le \ r$$

Choose t> 0 such that d (t x , 0) < ε. Then

$$d \ ((1 + t \ ) \ x, \ 0) \ \le \ r.$$

But, since (X, d) is strictly convex,

$$r = d \ (x, \ 0) < d \ ((1 + t) \ x, \ 0)$$

(from corollary 3.1 of the third chapter), which is a contradiction.

Then following theorem gives the structure of line segments in strictly convex linear metric spaces.

Theorem 2.2.2 [69] Let (X, d) be strictly convex and $r > 0$. Suppose $S[0, r] \neq \emptyset$ and y, z are distinct points of $B[0, r]$. Then

$$E = \{t \in R : ty + (1-t)z \in B[0,r]\}$$

is a compact convex subset of R.

Proof The convexity of E follows from that of $B[0, r]$

Clearly E is closed. Let $v = (y - z)$. Suppose E is not bounded above. Then $[0, \infty) \subset E$ so that

$$z + t\,v \in B[0, r] \text{ for all } t \in R^+$$

for any, $s \in (0, 1)$ and $t \in R^+$ we have

$$sz + tv = s(z + \frac{t}{s}v) + (1-s)0 \in B[0,r]$$

Since $B[0, r]$ is convex, $tv \in B[0, r]$ for any $t \in R^+$ and hence for any $t \in R$. Let $x \in S[0,r]$ (such a point exists by hypothesis). For any $x \in (0, 1)$ and $t \in R$ we have

$$sx + tv = sx + (1-s)(\frac{t}{1-s})v \in B[0, r]$$

Hence $x + tv \in B[0,r]$ for all $t \in R$.

In particular

$$x + v, \ x - v \in B[0, r]$$

Also

$$x + v \neq x - v$$

and $\quad x = \frac{1}{2}(x + v) + \frac{1}{2}(x-v)$

so from the strict convexity $x \in B(0, r)$, which is a contradiction. Therefore E is bounded above. Similarly it can be shown that E is bounded below. This completes the proof.

Note: Example 2.2.3 shows that the above result need not be valid if 'strict convexity' is replaced by ball convexity.

Corollary 2.2.1 [69] Let (X, d) be strictly convex and $r > 0$. Suppose $x \in X$, $S[0, r] \neq \emptyset$ and y, z are distinct points of $B[x, r]$. Then

$$E = \{t \in R : ty + (1-t)z \in B[x,r]\}$$

is a compact convex set.

26

Proof Since we can write E as

$$E = \{t \in R : t(y-x)+(1-t) \ (z- x) \in B[0,r]\},$$

the result follows from Theorem. 2.2.1

Corollary 2.2.2 [69] Let (x, d) be a strictly convex linear metric space. Then sup {d (t x, 0): t ∈ R} is invariant on X\{0}. In fact,

sup{d (t x, z): t ∈ R} is invariant on (X\{0}) x X .

Proof: Let u, v ∈ X and x, y ∈ X\{0} let

r = sup {d (t x, u) : t ∈ R}

and   s = sup {d(t y, v) : t ∈ R}

Suppose r < s then there exists a ∈ R such that

$$d(\alpha y, v) = r$$

Hence S [0, r] ≠ ∅. Also 0 and x are distinct points of B [u, r] and

$$\{t \in R : tx + (1-t) \ 0 \in B[u, r]\} = R$$

which is false in view of corollary 2.2.1. Hence r ≮ s similarly, it can be shown s ≮ r. Here r = s.

In Corollary 2.2.2, we have shown that in the presence of strict convexity, sup {d(t x, 0) : $t \in R$} is invariant on X\{0}.

In a strictly convex linear metric space, every half-ray emanating from the centre of a ball, passes through its surface, provided, of course, the surface is non-empty. This is the essence of

Corollary 2.2.3 [69] Let (x, d) be strictly convex, r > 0 and s[0,r] ≠ ∅. Suppose x, y ∈ x and y ≠ 0. Then

$$x+\alpha y \in S[x, r]$$

for some $\alpha \in R^+$

Proof Let z ∈ S[0,r]. Then, from corollary 3.1 (Chapter III) it follows that

sup {d(tz, 0)/ t ∈ R} > r.

Hence, by Corollary 2.2.3,

sup {d(ty, 0)/ t ∈ R} > r.

Consequently,

$$\exists s \ \alpha \in R^+$$

such that α y ∈ S [0, r]

27

so that

$$x + \alpha y \in S [x, r].$$

The above two results give the impression that strictly convex liear metric spaces behave like normed linear spaces.

The following two examples show that mere ball convexity does not guarantee the invariance of

$$\sup \{d (tx, 0) : t \in R\} \qquad \text{on } X \setminus \{0\}$$

In the first example we follow the technique used in the proof of the following result of Walter Rudin (Theorem 1.24 of [65]). "If X is a topological vector space with a countable local base, then there is a metric d on X such that

(a)   d is compatiable with the topology of X

(b)   The open balls centred at 0 are balanced, and

(c)   d is invariant i.e

$$d (x + z, y + z) = d (x, y) \text{ for x, y, z} \in X .$$

If, in addition, X is locally convex, then d can be choosen so as to satisfy (a), (b), (c) and also

(d)   all open balls are convex."

Example 2.2.7 [69]

Let $V_1 = \{(x, y) \in R^2 : |y| < \frac{1}{2}\}$, and

$V_n = \{(x, y) \in R^2 : (x^2 + y^2)^{1/2} < \frac{1}{2^n} \}$ for n= 2, 3 ...

Then

$$\{V_n : n = 1, 2, ...\}$$

is a balanced convex local base at the origin for the Euclidean topology on R^2. Also

$$V_{n+1} + V_{n+1} \subset V_n \text{for } n = 1, 2, ...$$

Let D be the set of rational numbers of the form

$$r = \sum_{n=1}^{\infty} C_n\,(r)\,\frac{1}{2^n}$$

where each of the digits $C_i(r)$ is 0 or 1 and only finitely many are 1. Define

$$A(r) = \begin{cases} R^2 & if \quad r \geq 1 \\ \sum_{n=1}^{\infty} C_n(r)V_n & if \quad r \in D. \end{cases}$$

Also define

$$f(x) = \inf \{n : x \in A(r)\} \text{ for } x \in R^2$$

and $d\,(x,\,y) = r\,(x - y)\,,\,(x \in R^2,\, y \in R^2)$

Then $(R^2,\,d)$ is a linear metric space, all of whose balls are convex. Further

$$\sup \{d\,(t(0,\,1),\,(0,\,0)\,) : t \in R\} = \frac{1}{2}$$

and

$$\sup \{d\,(t(1,\,0),\,(0,\,0)\,) : t \in R\} = 1$$

The metric d in this example can be explicitly expressed as follows:

Let $(x,\,y) \in R^2$. Then

$$d(x,\,y),\,(0,\,0)\,) = \begin{cases} \|(x,y)\| & if & \|(x,y)\| \leq \dfrac{1}{2} \\ \dfrac{1}{2} & if & \|(x,y)\| \geq \dfrac{1}{2} \text{ and } \|y\| \leq \dfrac{1}{2} \\ |y| & if & \dfrac{1}{2} \leq |y| \leq 1 \\ 1 & if & |y| \geq 1 \end{cases},$$

where $\|(x,\,y)\| = (x^2 + y^2)^{1/2}$

is the Euclidean norm. Here we have

$$S[0,r] = \begin{cases} \{(x,y) \in R^2 : \ \|x,y\| = r\} \ \text{if } 0<r<\dfrac{1}{2} \\[2ex] \{(x,y) \in R^2 : \ \|x,y\| \ge \dfrac{1}{2}, \|y\| \le \dfrac{1}{2}\} \ \text{if } r = \dfrac{1}{2} \\[2ex] \{(x,y) \in R^2 : \ |y| = r\} \qquad \text{if } \dfrac{1}{2}<r<1 \\[2ex] \{(x,y) \in R^2 : \ |y| \ge 1\} \qquad \text{if } \dfrac{1}{2}<r=1 \end{cases}$$

Example 2.2.8 [69] Define d on R^2 as follows:

$$d((x,y),\ (0,0)) \begin{cases} \|(x,y)\| & \text{if } \ \|(x,y)\| \le \dfrac{1}{2} \\[2ex] \dfrac{1}{2} & \text{if } \ \|(x,y)\| \ge \dfrac{1}{2} \text{ and } |y| < \dfrac{1}{2} \\[2ex] |y| & \text{if } \ |y| \ge \dfrac{1}{2} \end{cases}$$

where

$$\|(x,\ y)\| = (x^2 + y^2)^{1/2}$$

is the Euclidean norm. Then $(R^2,\ d)$ is a linear metric space with ball convexity. The metric nature of d follows immediately, if we observe that

$$|y| \le d\ ((x,\ y),\ (0,\ 0)\ \le \|x,\ y\|$$

for all $(x,\ y) \in R^2$

In example 2.2.4, the ' **superemum'** is finite in each direction whereas, here the **'superemum'** is finite in one direction and infinite in another. Infact,

$$\sup\{d(\ (\ t,\ 0),\ (0,\ 0)\ ) : t \in R\} = \frac{1}{2}$$

and

$$\sup\{d\ ((\ 0,\ t)\ ,\ (0,\ 0) : t \in R\} = \infty$$

Now we shall show that closed balls in a strictly convex linear metric space with non-empty surface are compact if the space is finite dimensional. We shall be using the following result of Walter Rudin (Theorem 1.28 (b) [65]) "If $\{x_n\}$ is a sequence in a metrizable topological vector space X and if $x_n \to 0$ as $n \to \infty$, then there are positive scalars γ_n such that

$$\gamma_n \to \infty \text{ and } \gamma_n x_n \to 0$$

Theorem 2.2.3 [69] Let (X, d) be strictly convex linear metric space and finite dimensional. If S {0, r] $\neq$ Ø then B [0, r] is compact.

Proof Since (X, d) is finite dimensional linear metric space, it is normable . Let $\|\cdot\|$ be a norm compatible with the topology on X. Since B [o,r] is closed, it is sufficient to show that it is norm bounded. Suppose that B [0, r] is unbounded in the norm. Then there exists a sequence $\{t_n\}$ of positive scalars and a sequence $\{x_n\}$ of vectors of unit norm such that $t_n \to \infty$ as $n \to \infty$ and $t_n x_n \in$ B [0, r) for all n. Since (X $\|\cdot\|$) is a finite dimensional normed linear space, $\{x_n\}$ has a convergent subsequence. We may suppose that $\{x_n\}$ is convergent with limit, say x_0. Then $\|x_0\| = 1$. Since the norm topology and d-topology are the same,

$$d(x_n - x_0, 0) \to 0 \text{ as } n \to \infty$$

Hence by the above theorem, there exists a sequence $\{\alpha_n\}$ of positive scalars such that

$$\alpha_n \to \infty \text{ and } d(\alpha_n(x_n - x_0), 0) \to 0 \text{ as } n \to \infty$$

Let $\beta_n = \min \{t_n, \alpha_n\}$. Then

$$\beta_n \to \infty \text{ as } n \to \infty.$$

For a positive integer m let $\varepsilon > 0$. There exists n > m such that

$$\beta_n > \beta_m \text{ and d } (\beta_n(x_n - x_0), 0) < \varepsilon.$$

Now

$$d(\beta_m x_0, 0) \le d(\beta_n x_0, 0)$$

31

$$\leq d \ (\beta_n(x_n - x_0), 0) \ + \ d(\beta_n \ x_n, \ 0)$$
$$\leq \ \varepsilon + d \ (t_n \ x_n, \ 0)$$
$$\leq \ \varepsilon + r.$$

This being true for each $\varepsilon > 0$, we have

$$d \ (\beta_m \ x_0, 0) \leq r$$

so that $\beta_m x_0 \in B[0,r]$ for each m. Since

$\beta_n \to \infty$ and B [0, r] is convex, it follows that

(1) sup $\{d \ (t \ x_0 \ , \ 0) : t \in R\} \leq r$

But, since S[0,r] $\neq \emptyset$, there exists y $\in$ S[0,r] and by theorem 3.1, it follows that

(2) sup $\{ \ d \ (t \ y \ , 0) : t \in R \ \} > r$

(1) and (2) contradict each other in view of corollary 2.2.2.

Hence B [0, r] is compact.

Next Lemma shows that a strictly convex finite dimensional linear metric space in the presence of an unbounded metric is totally complete, a notion introduced in [1]

Lemma 2.2.3 [69] A strictly convex finite dimensional linear metric space with an unbounded metric is totally complete.

Proof: Let (X, d) be strictly convex, finite dimensional linear metric space and d be unbounded. Let r > 0. Since d is unbounded, there exists y $\in$ X such that

d (y, 0) > r.

Hence by the continuity, there exists t $\in$] 0, 1 [such that d (t y, 0) = r so that S [0, r] $\neq \emptyset$. Therefore by Theorem 2.2.3, B [0, r] is compact. Hence, every closed ball and therefore every d-bounded closed set is compact.

In a linear metric space if the metric is additive along a half-ray emanating from the origin, then it is a norm along the line determined by the half-ray, More generally, we have the following result, the proof of which is immediate.

Lemma 2.2.4 [69] Suppose $x_0, y_0 \in X$ are such that

(1) $d(x, y) = d(x, z) + d(z, y) \ \forall \in [x, y]$

whenever $x, y \in [x_0, y_0]$. Then

$d(tx_0, ty_0) = t \ d(x_0, y_0)$ for all $t \in [0, 1]$.

The above result need not be true, even when (X, d) is strictly convex if (1) is replaced by

$$d(x_0, y_0) = d(x_0, z) + d(z, y_0) \ \forall z \in (x_0, y_0)$$

as the following example shows:

Example 2.2.9 [69] Define $f: R^+ \to R^+$ by

$$f(t) = \begin{cases} \dfrac{4}{3} t & \text{if} & 0 \le t \le \dfrac{1}{4} \\[2mm] \dfrac{2}{3} t + \dfrac{1}{6} & \text{if} & \dfrac{1}{4} \le t \le \dfrac{3}{4} \\[2mm] \dfrac{2}{3} t + \dfrac{1}{3} & \text{if} & t \ge 1 \end{cases}$$

Then (R, d), where

$$d(x, y) = f(|x - y|) \ \forall x, y \in R,$$

is a strictly convex linear metric space. Clearly

$$d(0, 1) = d(0, t) + d(t, 1) \ \forall t \in [0, 1]$$

but

$$d(0, t) = td(0, 1) \ \forall t \in [0, 1]$$

is not true.

The following example shows that the distance between two points can be the sum of their distances from an intermediate point but at the same time it may not be so for every intermediate point, even in a strictly convex linear metric space.

Example **2.2.10** [69] Define $f : R^+ \to R^+$ by

$$f(t) = \begin{cases} \frac{2t}{1+t} & if \quad 0 \le t \le 1 \\ t & if \quad t \ge 1 \end{cases}$$

Then (R, d), where

$$d(x, y) = f(|x - y| \ \forall \ x, y \in R,$$

is a strictly convex linear metric space. We have

$$d(3, 4) + d(4, 5) = d(3, 5) \ne d(3, \frac{7}{2}) + d(\frac{7}{2}, 5)$$

Now we show that strict convexity is weaker than

property (P) but stronger than the property (P_1). This is the essence of our next theorem.

Theorem 2.2.4 [54] Let (X, d) be a linear metric space, we have:

(i)   If (X, d) has property (P) then it is strictly convex.

(ii)   If (X, d) is strictly convex then it has property (P_1).

Proof (1) Suppose (X, d) is not strictly convex. Then by lemma 3.2 (of chapter III) there exists an r > 0 and distinct points x and y such that

$$d (x, 0) = d (y, 0) = r$$

and   B [0, r] $\cap$] x, y [= $\emptyset$.

consider the compact line segment [x, y]. This set is proximinal let $f : E \to [x, y]$ be the nearest point mapping then

$$f (0) = x, f (0) = y. \text{ Consider}$$

$$d (x, y) = d (f (0) , f (0)) \le d (0, 0) = 0$$

[By property (P)] and so x = y, a contradiction.

(ii)   If d (x + z, 0) < d (x, 0) and
        2d (y, x) ≤ d (x, 0) – d (x + z, 0)

then

$$d (y + z, 0) \leq d (x + z, 0) + d (y, x) \leq d (y, 0)$$

Thus property (P_1) is satisfied if

$$b = [d ((x,0) - d (x + z, 0)]/2 \text{ and } C = 1.$$

If d (x + z, 0) = d x, 0)

then by the strict convexity,

$$d (x + \frac{z}{2}, 0) = d (\frac{x + z + x}{2}, 0) < d (x ,0)$$

and so property (P_1) is satisfied if

$$b = [d \{x, 0) = d(x + \frac{z}{2},0)]/2 \text{ and } C = \frac{1}{2} \text{ as}$$

$$d (y + \frac{z}{2}, 0) \leq d(y, x) + d (x + \frac{z}{2}, 0)$$

$$= d (y, x) + d (x, 0) - 2 b$$

$$\leq x (x, 0) - b$$

$$\leq d (y, 0).$$

Theorem 2.2.5 [1] A complete convex set K in a linear metric space (X, d) satisfying the property (P) is Chebyshev.

Proof :   Let g ε X and

$$r = \inf \{d (x, g) : x \in K\}$$

By definition of infimum there is a sequence $< x_n >$ in K such that

35

$$\lim_{n \to \infty} d(x_n, g) = \inf \{d(x, g): x \in K\}$$

By property (P) we have $< x_{n_k} >$ in K. K being complete,

$$< x_{n_k} > \longrightarrow x \in K$$

and consequently

$$d(x^*, g) \geq r.$$

Also $d(x^*, g) \leq d(x^*, x_{n_k}) + d(x_{n_k}, g)$

implies

$$d(x^*, g) \leq r.$$

Hence

$$d(x^*, g) = r$$

Now, if possible

$$x_1^*, x_2^*, \in K$$

be such that

$$d(x_1^*, g) = d(x_2^*, g) = r.$$

Consider the sequence $< x_n >$ defined as

$$x_n = \begin{cases} x_1^* & \text{if } n \text{ is odd} \\ x_2^* & \text{if } n \text{ is even.} \end{cases}$$

Then $\lim_{n \to \infty} d(x_n, g) = d(x_1^*, g) = d(x_2^*, g) = r = \inf \{d(x, g) : x \in K\}.$

By property (P), $< x_n >$ has a Cauchy sequence $< x_{n_k} >$ and therefore for a given $\varepsilon > 0$, there exists a positive integer N such that

$$d(x_{n_k}, x_{m_k}) < \varepsilon \text{ for all } n_k, m_k \geq N,$$

i.e. $d(x_1^*, x_2^*) < \varepsilon$

ε being arbitrary, $x_1^* = x_2^*$.

--- X ---

CHAPTER III

CHARACTERIZATIONS OF STRICTLY CONVEX LINEAR METRIC SPACES

This chapter is devoted to study some characterizations of strictly convex linear metric spaces. First of all, we give the necesssary and sufficient condition for the real line to be strcitly convex (Theorem 3.1). Theorem 3.2, which characterizes strictly convex linear metric spaces in terms of best approximation, shows that the converse of the unicity theorem "Every Convex Proximinal set in a Strictly Convex Linear Metric Space is Chebyshev" is also true. Theorem 3.3 tells that a linear metric space (X, d) is strictly convex if and only if all convex subsets of X are semi-Chebyshev. Theorem 3.4 shows that a linear metric space is strictly convex iff every non empty locally compact closed convex set of it is a Chebyshev set.

The following theorem gives necessary and sufficient condition for the real line to be strictly convex.

Theorem 3.1 [67] Let d be an invariant linear metric on R and let $f : R^+ \to R^+$ be defined by $f(t) = d(t, 0)$. Then (R, d) is strictly convex if an only if f is strictly increasing.

Proof Suppose (R, d) is strictly convex. Let $0 \le s < t$. Since the ball $B[0, f(t)]$ is convex (From Theorem 2.1.2), we must have

$$f(s) = d(s, 0) \le f(t)$$

Thus f is increasing on R^+. Now by strict convexity of

$$(R,d), \; f\left(\frac{(s+t)}{2}\right) = d\left(\frac{s+t}{2}, 0\right) < f(t)$$

so that

$$f(s) < f\left(\frac{(s+t)}{2}\right) < f(t).$$

Thus f is strictly increasing on R^+.

Conversely, suppose that f is strictly increasing on R^+. Let x, y be distinct points of R. Then

$$d\left(\frac{(x+y)}{2}, 0\right) = f\left(\frac{|x+y|}{2}\right)$$

$$\le f\left(\frac{|x|+|y|}{2}\right)$$

$$< \quad \max \{f(|x|),\ f(|y|)\}\ (\text{if } y \neq -x)$$
$$= \max \{d(x, 0),\ d(y, 0)\}$$

If $y = -x$, then

$$d\left(\frac{x+y}{2}, 0\right) \ = 0 < \max\{d(x, 0),\ d(y, 0)\}$$

Hence (R, d) is strictly convex.

The following is an easy consequence of Theorem 3.1.

Corollary 3.1 [67] Let (X, d) be a strictly convex linear metric space and $0 \neq x \in X$. Define $f_x : R^+ \to R^+$ by

$$f_x(t) = d(t x, 0)$$

Then $f_x(0) = 0$, f_x is continous and strictly increasing on R^+.

Proof Given that (X, d) is strictly convex. Let $0 \leq s < t$.

Since the ball $B[0, f_x(t)\}$ is convex, we must have

$$f_x(s) = d(s x, 0) \leq f_x(t).$$

Thus f_x is increasing on R^+. Now by strict convexity of (X, d)

$$f_x\left(\frac{(s+t)}{2}\right)\ = d\left(\frac{(s+t)}{2} x, 0\right) < f_x(t)$$

so that

$$f_x(s) < f_x\left(\frac{(s+t)}{2}\right) < f_x(t)$$

Thus f_x is strictly increasing on R^+.

Now $f_x(0) = d(0 x, 0) = d(0, 0) = 0$. f_x is always continuous as distance function is always continuous.

The following theorem on unicity of best approximation was proved in chapter II (Theorem 2.1.2) : Every Convex proximinal set in a strictly convex linear metric space is Chebyshev. We shall show that converse of this theorem is also true. In order to show this, we establish two lemmas.

Lemma 3.1 [42] In a linear metric space (X , d) the line segment
$$[x, y] = \{ t x + (1+ (1-t) y : 0 \leq t \leq 1\}$$

is compact and convex.

Proof The proof follows from well known result (cf [5], p.9) "the Convex hull of the union of finitely many compact convex subsets of a toplogical linear space is compact."

There are two or more equivalent forms for the notion of strict convexity in normed linear spaces which can be generalized to linear metric spaces and in the following lemma, we establish them.

Lemma 3.2 [51] For a linear metric space (X, d) the following are equivalent:

(i)   (X , d) is strictly convex

(ii)   Balls in (X, d) are convex and S $[0, r] = \{ x \in X : d (x, 0) = r\}$ does not contain any line segment for $r > 0$.

(iii)   x , y $\in$ B $[0, r] \Rightarrow$] x, y [$\subset$B (0, r).

Proof (i) => (ii) "Since the space is strictly convex, the balls in it are convex. The second part is immediate from the definition.

(ii)=> (iii) Let x, y $\in$B [0, r], x $\neq$ y, r > 0 let u $\in$] x, y [, then there exists t, $0 < t < 1$ such that

$$u = t x + (1-t) y.$$

Choose $\delta > 0$ such that

$$0 < t - \delta < t < + \delta < 1.$$

Let

$$v = (t - \delta) x + (1 - t + \delta) y, \quad w = (t + \delta) x + (1 - t - \delta) y.$$

Since B [0, r] is convex v, w $\in$ B [0, r] and consequently, by strict convexity,

$$d (u,0) = d (t x + (1-t) y, 0)$$

$$= d \left(\frac{v + w}{2}, 0\right) < r.$$

(iii) => (i) is obvious.

The following theorem which characterizes strictly convex linear metric spaces in terms of best approximation shows that the converse of unicity theorem (Theorem 2.1.2) is also true.

Theorem 3.2 [51] In a linear metric space (X, d) the following statements are equivalent :

(i)   X is strictly convex

(ii)   For each convex set S and distinct point x and y of

     S, $S_x \cap S_y = \emptyset$

     Where $S_y = \{ x \in X : d (x, y) + d (x, S) \}$

(iii)   Whenever a convex set is proximinal it is uniquely proximinal.

Proof (i) => (ii) Let $S_x \cap S_y \neq \emptyset$. Let

$$z \in S_x \cap S_y => d(z, x) = d(z, y) = d(z, S)$$

Strict convexity of the space implies

$$d\left(\frac{(z-x)+(z-y)}{2}, 0\right) < d(z, S)$$

i.e. $d(z, \frac{x+y}{2}) < d(z, S)$

which is a contradiction, for $\frac{x+y}{2} \in S$.

(ii) => (iii) Let a convex set S be proximinal and p be any point of X Because S is proximinal, there exists $x \in S$ such that $p \in S_x$. Let, if possible, $y \neq x$ be also nearest to p.

Then $p \in S_y$ and so $p \in S_x \cap S_y$, $x \neq y$ which is a contradiction.

Thus x is the only point of S closest to p i.e. S is Chebyshev.

(iii)=>(i) Suppose X is not strictly convex. Then by lemma 3.2, there exists $r > 0$ and distinct points x and y such that $d(x, 0) = d(y, 0) = r$ and $B(0, r) \cap]x, y[= \emptyset$.

Now consider the compact convex line segment [x, y]. This is proximinal (a compact subset of a metric space is proximinal, cf [5], p. 123) but not uniquely proximinal since for the point 0 of X there are atleast two nearest points (namely x, y) which contradicts (iii). Hence the space must be strictly convex.

The following theorem gives another characterization of strictly convex linear metric spaces in terms of best approximation.

Theorem 3.3 [54] A linear metric space (X, d) is strictly convex iff convex subsets of X are semi-Chebyshev.

Proof : Let (X, d) be strictly convex and G be a convex subset of X. Suppose there exists some $x \in X \backslash G$ which has two distinct best approximations in G say g_1 and g_2 i.e. $d(x, g_1) = d(x, g_2) = d(x, G)$. Then by the strict convexity of d,

$$d\left(x, \frac{g_1 + g_2}{2}\right) < d(x, G),$$

a contradiction as $\frac{g_1 + g_2}{2} \in G$. Therefore G must be semi-Chebyshev.

41

Conversely, suppose all convex subsets of the linear metric space (X, d) are semi-Chebyshev. Suppose (X, d) is not strictly convex. Then by lemma 3.2, there exists an $r > 0$ and distinct points x, y $\in X$ such that

$\qquad$ d (x, 0) = d (y, 0) = r

and

$\qquad$ B (0, r) $\cap$ } x, y [= $\emptyset$.

Consider the convex line segment [z, y]. It is not semi-Chebyshev since for the point 0 of X there are two distinct best approximations (x and y), a contradiction.

Remark : Replacing the line segment [x, y] by the real one dimensional subspace G = {α (y $-$x) : $-\infty < \alpha < \infty$} in the second part of the proof of the above theorem we can see that G is not semi-Chebyshev as for the element $-$x $\in$ X, both 0 an y-x are best approximations in G and hence it follows that a linear metric space (X, d) is strictly convex if and only if all linear subspaces of X are semi-Chebyshev.

Now we give another characterization of strictly convex linear metric space given by A. I. Vasil'ev in [18]. Before this, we discuss some lemmas to be used in the characterization.

Lemma 3.3 [93] Suppose X is a linear metric space. Then the following conditions are equivalent :

(1) The Space X is strictly convex

(2) If x, y $\in$X, x $\neq$ y, 0 < α < 1, then

$\qquad$ | (1-α) x + αy | < max ($|x|$, $|y|$).

(3) Every ball B [0, r] in X is convex and no sphere S[0, r] contains segments.

(4) The quasinorm of X is strictly monotone and all balls B[0, r] are strictly convex.

(5) For any r > 0, the ball B[0, r] is strictly convex and coincides with the closure $\overline{B}$ (0, r) of the open ball B (0, r).

Proof : We first note that (1) is equivalent to the following condition

(1) If x, y $\in$ X, x$\neq$ y and $|x| \leq$ r, $|y| \leq$r, 0 < α < 1,

then $|(1 - \alpha x) + \alpha y| < r$.

Indeed, we deduced from (1) in the usual way that each B[0, r] is convex, which, together with (1) yields (1') , and the implication (1') => (1) is obvious.

$\qquad$ Since (1') $\Leftrightarrow$ (2), it follows that (1) $\Leftrightarrow$ (2)

We will prove that (1) => (5) => (4) => (3) => (1).

<u>(1)=> (5)</u> Clearly, B [0, r] is strictly convex for an arbitrary r > 0.   If x ∈B [0, r]/
$\overline{B}$ (0, r) then there exists y ∈] 0, x [such that

$$[y, x] \cap B (0,r) = \emptyset$$

and therefore $\left|\dfrac{(x+ y)}{2}\right| = r,$ which contradicts (1)

Consequently, B[0,r] = $\overline{B}$ (0,r).

<u>(5)=>(4)</u> suppose x ≠ 0, 0< $\alpha'< \beta$ and $r = |\beta x|$, Then

$$B (0, r) = \underset{s<r}{U} B[0, s]$$

is convex, $\beta x \in \overline{B}(0,r), 0 \in B(0,r) = \text{int } B (O,r), \alpha' x \in B(0,\beta x).$

So $\alpha'x \in B(0,\beta x)$ i.e. $|\alpha'x| < |\beta x|.$

<u>(4) =>(3)</u> Suppose (4) is satisfied, but not (3) i.e. there exists r >0 and two distinct
points x, y∈ such that

$$S[0, r] \supset [x, y]$$

Take z ∈] x, y [. Since B[0,r] is strictly convex, it follows that z ∈ int B [0,r],
hence there exists $\alpha' > 1$ such that

$$|\alpha'z| \le r = |z|,$$

which contradicts the strict monotonicity of the quasinorm.

<u>(3) => (1)</u> Suppose x, y ∈B[0,r], x ≠ y, z = $\dfrac{(x+y)}{2}$. Since

B[0, r] is convex, we have
[x, y] ∈ B[0,r].
and since S[0,r] contains no segments, each of [x, z] and [z, y] meets B(0,r), which
is obviously convex. Therefore z∈ B(0,r).

Hence the lemma is complete.

Lemma 3.4 [93] Suppose E is a closed bounded subset of a finite dimensional linear
metric space. If each face of E is singleton, then ∂(conv E) ⊆ E.

Proof For the proof of this lemma, we refer to [42, Theorem 9].

Leema 3.5 [93] Suppose X is a linear metric space and

$$I =] 0, \alpha],] 0, \alpha [\text{ or }] 0, \infty [.$$

Then the following assertions are equivalent:

(1)   For any r∈ I, B[0,r] is convex, B[0,r] ≠ X and
        B[0,r] = $\overline{B}$ (0,r).

(2)   For every $r \in I$, conv$B[0,r] \neq X$ and $\partial[0,r]) \subset \overline{B}(0,r)$

Proof For proof of this lemma, we refer to [92]

Lemma 3.6 [93] Suppose each closed half space in a finite-dimensional linear metric space X is a uniqueness set. Then each bounded ball $B[0,r] \subset X$ is strictly convex and coincides with

$\overline{B}$ (0,r)

Proof We shall first show that (1) (conv $B[0,r]$) $\subset B(0,r)$ for every bounded ball $B[0,r] \subset X$.

Suppose $B[0,r]$ is a bounded ball, H is a hyperplane supporting $B[0,r]$ with the equation $h(x) = b$ (b>0),

$H = \{X \in X; h(X) \geq b\}$ and $\Gamma = B[0,r] \cap H$. Obviously $d(0, x) \geq r$ for any $X \in H$ and $\emptyset \neq \Gamma = B[0,r] \cap H$. Consequently, $\Gamma = P_H$ (0), hence (since H is a uniqueness set) Γ is singleton. Thus, any face of the ball $B[0,r]$ is a singleton. Assertion (1) now follows from lemma 3.4. we will prove that (2) (conv $B[0,r]$) $\subset \overline{B}$ (0,r) for any bounded ball $B[0, r] \subset x$.

Assume the contrary. Then for some bounded ball $B[0, r]$ there exists a point $w = \partial(\text{con } B[0,r]) \backslash \overline{B}$ (0,r). Let $f(x) = c$ be the equation of a hyperplane supporting conv $B[0,r]$ at the point w (c >0). Obviously ,

$\quad$ $f(w) \geq \sup \{ f(x) : x \in \overline{B}$ (0,r) $= : a$

$\quad$ let $F = \{x \in X : f(x) \geq a \}$.

Since $\overline{B}$ (0, r) is compact, there exists $x_0 \in \overline{B}$ (0, r) such that f(x_0)= a. It is clear that $x_0 \neq w$, d (0, x_0) = r and d (0, x) $\geq$ r and therefore in view of (1), d (0, w) = r. Thus

$\quad$ d (0, F) = d (0, x_0) = d (0, w),

so that F is not a uniqueness set. This contradiction to the hypothesis of the lemma proves (2). Let $I = \{ V_r, r > 0\}$ Then $I =]0, \alpha]$, $] 0, \alpha[$ or $] 0, \infty [$ and conv B $[0, r]$ $\neq X$ for each r $\in$ I. It follows from (2) and lemma 3.5 that every bounded ball B $[0, r]$ is convex and coincides with $\overline{B}$ (0, r). From this fact and the condition that every closed hair space is a uniqueness set, we now obtain the strict convexity of bounded balls B $[0, r]$ and hence the proof of the lemma is proved.

Lemma 3.7 [93] Suppose M is a closed convex symmetrical subset of a finite dimensional linear metric space X. If M is unbounded and $M \neq X$, then ∂M contains a straight line.

Proof Suppose $0 \in M$, it follows (see [35, p. 343]) that M contains a ray of the form $\{x : \alpha \geq 0\}$, $x \neq 0$, and since M is symmetric $M \supset \{ \alpha x : \alpha \in R\} = L$. Take $z \in \partial M$ and put $P = L + z$. Since M is convex and closed, it follows easily that $P \subset M$, using the property] y, v [$\subset$ int M, we see that if $y \in$ int M and $v \in M$, then $P \subset \partial M$.

Lemma 3.8 [93] A strictly convex proper ball in a linear metric space contains no rays.

Proof Assume that a strictly convex proper ball B [0, r] in a linear metric space X contains a ray. Since B [0, r] is convex and closed it contains a ray of the form

$$\{\alpha x : \alpha \geq 0\} = \ell, x \in X \setminus \{0\}.$$

Take a two-dimensional subspace L of X such that $\ell \subset L$ and $V_r \cap L \neq L$. By lemma 3.6, the boundary in L of the set B [0, r] $\cap$ L containing some straight line P, hence $\partial B[0,r] \supset P$, which contradicts the strict convexity of B [0, r]. Hence the lemma is proved.

It is well known (see [41], p. 347) that in a normed linear space, every nonempty closed convex set which is locally compact in the weak topology is an existence set. For a linear metric space we have

Lemma 3.9 [93] Suppose that every ball B [0, r] in a linear metric space X is convex. Then the following conditions are equivalent:

(1) Every nonempty closed convex set in X which is locally compact in the weak topology is an existence set.

(2) No proper ball B[0, r] contains a ray of the form $\{\alpha x : \alpha \geq 0\}$, $x \in X \setminus \{0\}$.

Proof (1) => (2) Assume that (2) does not hold, i.e. there exists a proper ball B [0, r] containing a ray

$$\{\alpha x : \alpha \geq 0\}, x \in X \setminus \{0\}.$$

Since B [0, r] is symmetric B [0, r] $\supset \{\alpha x : \alpha \in R\} = : R x$.

45

Take two dimensional subspace L of X such that L $\supset$ Rx and B $[0, r] \cap L \neq L$. We view L as a linear metric space with the metric induced from X. Take y $\in$ L\Rx. If we associate to each point $a_1 x + a_2 y \in$ L and define a matric

d' on R^2 for $(a_1, a_2) \in R^2$

d' (a, b) = d $(a_1 x + a_2 y, b_1 x, + b_2 y)$

$(a = (a_1, a_2) \in R^2, b = (b_1, + b_2) \in R^2)$,

we can identify the liner metric space L with the linear metric space (R^2, d). The implication (1) $\Rightarrow$ (2) will be proved if (R^2, d') contains a nonempty closed convex set that is not an existence set. The ball $B'[0, r] = \{a \in R^2, d'0, a) < r\}$ where $0' = (0, 0)$, contains the straight line

$\{a, 0) \in R^2 : \alpha \in R\}$

and is a symmetric closed convex set different from R^2. It follows easily that $B'[0, r]$ is a closed strip of the form

$\{(a, u) \in R^2 : |\mu| < \alpha' \}, \geq 0$

Now consider the set

M = $\{ (\alpha, u + \alpha') : \alpha > 0, u > 0, \alpha u \geq 1\}$

Since for any $\varepsilon > 0$ there exists $u > 0$ such that

$(0, u) \in B'[0, \varepsilon]$

and therefore for $\alpha \geq \dfrac{1}{\mu}$ we have the relations

$(\alpha, u + \alpha') \in M$

and

$(\alpha, u + \alpha') \in \{ (\alpha, \alpha') \in R^2 : \alpha \in R \} + B'[0, \varepsilon] \subset B[0, r + \varepsilon]$

49

it follows that $\varepsilon > 0$, B $[0, r + \varepsilon] \cap M = \emptyset$ for any $\varepsilon > 0$.

Using the equality

$B'[0, r] \cap M = \emptyset$

we see that the nonempty closed convex subset M of the space (R^2, d') is not an existence set. Hence (1) => (2) is proved.

$(2) \Rightarrow (1)$ Suppose M is nonempty closed convex set in X that is locally compact in the weak topology. We must show that

$P_M(z) \neq \phi \qquad$ for $z \in X \backslash M$ where

$P_M(z) = \{ y \in M : d(z, y) = d(z, M) \}, z \in X.$

In view of the invariance of the metric with translation, we may assume without loss of generality that $z = 0$.

Let $r_0 = \sup \{ |x| : x \in X \} \leq \infty, s = d(0, M)$.

If $s = r_0$, then $d(0, x) \geq r_0$ for $x \in M$, hence $d(0, x) = r_0$, so that $P_M(z) \neq \phi$. Suppose $s < r_0$ for $r \in (s, r_0)$, $B[0, r] \cap M$ is nonempty and weakly compact (being a convex weakly closed set that is locally compact in the weak topology and contains no rays), consequently

$$P_M(0) = \bigcap_{s < r < r_0} (B[0, r] \cap M) \neq \phi$$

This proves the lemma.

Lemma 3.10 [93] In a strictly convex linear metric space, every nonempty closed convex set which is locally compact in the weak topology is a Chebyshev set.

Proof Suppose M is a subset of a strictly convex linear metric space X satisfying the given conditions. Since each ball in X is strictly convex (see lemma 3.3), it follows from lemma 3.8 and 3.9 that M is an existence set. Since X is strictly convex, it follows that M is a Chebyshev set.

Lemma 3.11 [93] Suppose X is a linear metric space with monotone quasinorm and dim $X \geq 2$. Then the following conditions are equivalent:

(1)   Every subspace of X is a uniqueness set.

(2)   Every one-dimensional subspace of X is a uniqueness set.

(3)   Every finite-dimensional subspace of X is a Chebyshev set.

(4)   Every one-dimensional subspace of X is a Chebyshev set.

(5)   The quasinorm of X is strictly convex and all balls b [0, r] are strictly convex.

Proof : For its proof, we refer to [6, p. 189].

The following theorem shows that a linear metric space is S.C. iff every nonempty locally compact closed convex set of it is a Chebyshev set.

Theorem 3.4 [93] Suppose X is a linear metric space. Then the following conditions are equivalent:

(A)   Every non-empty locally compact closed convex set in X is a Chebyshev set.

(B)   Every segment in X is a Chebyshev set.

47

(C_n) Every closed n-dimensional half plane in X (n fixed, $1 \leq \dim$
    $X \leq \infty$) is a Chebyshev set.
(C)   The space X is strictly convex.

Proof It suffices to show that (A) => (B) => (D) => (A) => (C_n)=>(D).
    The implications (A) => (B) and (A) => (C_n) are obvious.

(B) => (D) Suppose that (B) holds. For an arbitrary r > 0 we will prove the strict convexity of the ball B [0, r] and the equality B [0, r] = $\overline{B}$ (0, r). Note that
(1) If z, y $\in$ B [0, r] , z $\neq$ y, then] z, y [$\cap$B[0, r] $\neq \emptyset$.

Otherwise we would have z, y $\in$ S [0, r] , d (0, [z, y])
= d (0, z) = d (0, y) and therefore [z, y] would not be a Chebysheve set. Using (1) and the fact B [0, r] is closed, we obtain in the standard way that B [0, r] is convex. In view of (B), the boundary of B [0, r] is convex. In view of (B), the boundary of B [0, r] (which obviously lies in S[0, r]) contains no segments. Thus, B [0, r] is strictly convex. Also,

    B [0, r] = $\overline{B}$(0, r),

since if x $\in$B [0, r]\ $\overline{B}$ (0, r) then there exists z$\in$ (o, x) such that [z, x] $\subset$S [0, r] , which contradicts (B).

(D) => (A) Indeed, since a convex locally compact subset of a seperated locally convex space is locally compact in the weak topology (see[7]) , the assertion (D) => (A) is a special case of lemma 3.10.

(C_n) => (D) Suppose (C_n) holds. Assume 1) dim X = n.

By lemma 3.6 each bounded ball B[0, r] is strictly convex and coincides with $\overline{B}$ (0, r). Therefore, it is sufficient to show that if B [0, r] is an unbounded ball then
    B [0, r] = B (0, r) = X.
Suppose there exist unbounded balls in X and
    r = inf {r > 0 : B [0, r] is unbounded}.
obviously r' > 0 .Take some basis { x_1, x_2, ... x_n } in X and define a norm $\| \cdot \|$ on X by putting $\| x \|$ = max { $|a_1|$: $i = 1,\ 2,...,n$} for

$x = \alpha_1 x_1 + \alpha_2 x_2 + ... + \alpha_n x_n$. For δ > 0 put

N_δ = $\{x \in X : \|x\| < \delta\}$,

P_i^δ = $\{\alpha_1 x_1 + \cdots + \alpha_n x_n$: $\alpha_1,...,\alpha_n \in R,\ \alpha_i \geq -\delta\}$

P_{n+1}^δ = $\{\alpha_1 x_1 + \cdots + \alpha_n x_n$: $\alpha_1,...,\alpha_n \in R,\ \alpha_i \leq -\delta\}$

where i = 1, 2 , ..., n. Note that

$$N_\delta = X \setminus \bigcup_{j=1}^{2n} P_j^\delta$$

We will prove that B $(0, r')$ is unbounded. Assume that B $(0, r')$ is bounded. Then B $[0, r']$ is also bounded (Otherwise, taking $\delta > 0$, for which B $(0, r')$

$\subset X \setminus \bigcup_{j=1}^{2n} P_j^\delta$, we see that for some j = 1, 2, ..., 2n the half space P_j^δ contains at least

two distinct points of S $[0, r']$ and, therefore, is not a uniqueness set).

Thus, there exists $\delta > 0$ such that

$$B[0,r'] \subset X \setminus \bigcup_{j=1}^{2n} P_j^\delta$$

For any j=1, 2 , ..., 2 n we have d $(0, P_j^\delta) > r'$, Indeed, if for some

$j = 1, 2, ..., 2n$ this is not so, then, snce the half space P_j is a Chebyshev set, we have

$$B\,[0,\,r']\cap P_j^\delta \neq \phi,$$

which contradicts the inclusion

$$B\,[0,\,r'] \subset X \setminus \bigcup_{j=1}^{2n} P_j^\delta.$$

Now take s such that

$$r' < s < d\,(0,\,P_j^\delta)$$

for each j = 1, 2 , ..., 2n. Since s < d $(0, P_j^\delta)$, it follows that

$$B\,[0,\,s]\cap P_j^\delta = \phi\ (\,j = 1,2,\,...,\,2n)$$

and therefore

$$B\,[0,\,s] \subset X \setminus \bigcup_{j=1}^{2n} P_j^\delta$$

Thus, B $[0, s]$ is bounded, which together with the inequality s > r', contradicts the definition of r'. This proves that B $(0, r')$ is an unbounded symmetric convex set. It is now clear that B $(0, r')= X$ if n = 1 suppose n > 1 and B $(0, r') \neq X$. Then $\overline{B}\,(0, r') \neq X$, and by lemma 3.7. $\partial \overline{B}\,(0, r')$ contains some straight line P $\subset$ S[0,

r']. By the Hahn-Banach Theorem, there exists a hyperplane H containing P and not meeting B(0, r').

Let $h(y) = c$ be the equation of this hyperplane $(c > 0)$,

and let $y \in H$, and since $P \subset H \cap S [0,r']$, we have

$$d (0, H) = r' = d (0, p) \quad \text{for any } p \in P.$$

This contradicts the fact that H is a Chebyshev set. Consequently B $(0, r') = X$ and for any $r > r'$,

B [0, n] = B (0, r) = X.

Thus, if dim X = n, then $(C_n) \Rightarrow$ (D).

Suppose 2) dim X $>$ n $= 1$. We will show that (2) the quasinorm of X is monotone, Suppose

$$z \in X \backslash \{0\}, 0 < \alpha < \beta.$$

Put $r = |\beta z|, L = \{\alpha' z, \alpha' \in R\}$. We wil view L as a linear metric space (wth metric induced from X). Since L satisfies (C_1), it follows from what has just been proved that $L \cap B[0, r] = \{x \in L : |x| \le r\}$ is convex. Consequently,

$$|\alpha z| \le r = |\beta z|.$$

Thus, (2) is proved.

We now observe that (3) every one-dimensional subspace of X is a uniqueness set.

Otherwise, there exist a straight line and points x, y $\in$ P $(x \ne y)$ such that d (0, P) = d (0, x) = d (0, y), thus, for $\ell = \{ x + \alpha' (y-x) : \alpha' \ge 0\}$ we have d (0, ℓ) = d (0, x) = d (0, y), which is not possible, since in view of $(C_n) = (C_1)$ the ray ℓ is a Chebyshev set. It follows from (2) and (3) (see lemma 3.11) that X is strictly convex. Thus, the implication $(C_n) \Rightarrow$ (D) is proved in case 2).

There remains : 3) dim X$>$ n $>$ 1. By what was proved above (case 1), every n-dimensional subspace of X is strictly convex from which it obviously follows that X itself is strictly convex. So $(C_n) \Rightarrow$ (D) and hence the theorem gets proved.

--- X ---

CHAPTER- IV

ON BEST APPROXIAMTION AND METRIC PROJECTIONS

For a given point x and a given set G of a metric space (X,d), a point $g_0 \in G$ satisfying d (x, g)=inf {d(x,g):g∈G} is called a best approximation to x in G and the map which takes each point x∈X to set of its best approximation map or the metric projection of X onto G. This chapter dealing with best approximation and metric projections, has been divided into two sections. First section is concerned with best approximation in pseudo strictly convex metric linear spaces and second section with multi-valued metric projections in convex spaces.

The notion of pseudo strict convexity in metric linear spaces was introduced and discussed by K.P.R. Sastry and S.V.R. Naidu in [69] and [71]. It was shown by Paul C.Kainen et al [38] that the existence of a continuous best approximation in a strictly convex normed linear space X and taking values in a suitable subset M of X implies that M has the unique best approximation property. In the first section of this chapter, we extend this result of Paul C.Kainen et al to pseudo strictly convex metric linear spaces.

S.B.Stockin [89] provod that if U_M= {x∈X: Card P_M (x) ≤1} then U_M =X for every subset M of X iff X is a strictly convex normed linear space. This result was extended to strictly convex metric spaces by T.D. Narang [49]. A question that arises is what happens in spaces which are not strictly convex? To answer this, we have discussed the characterization of multi-valued metric projection P_M in spaces which are not strictly convex in the second section. For normed linear spaces this result was proved by Ioan Serb in [72]. We have also proved that for a non-void proper subset M of a complete convex metric linear space (X,d) P_M cannot be a countably multi-valued metric projection. We have given a characterization of the semi-metric linear spaces in terms of finitely multi-valued metric projections. In [73], it was proved that if M is a strongly proximinal subset of a Banach space X, then card P_M (x)≥c for every x∈X\M, and the completeness of the space is essential for the validity of the result. In [74], the same result was proved for complete metrizable locally convex spaces i.e. in Frechet spaces. In this section, we have proved that for a strongly proximinal set M in a complete convex metric space (X,d) card $P_M(x) \geq c$ for all $x \in X \setminus M$.

4.1 Best Approximation

This section deals with best approximation in pseudo strictly convex metric linear spaces. To begin with , we recall a few definitions .

Definition 4.1.1 [90] A metric linear space (X,d) is said to be **convex** if for x, y $\in X$, $\lambda \in [0,1]$

$$d(u, \lambda x + (1-\lambda)y) \leq \lambda d(u,x) + (1-\lambda)d(u,y)$$

for all u $\in X$.

Definition 4.1.2 [75] A metric linear space (X,d) is said to be **pseudo strictly convex (P.S.C.)** if given x $\neq 0$, y $\neq 0$, d(x+y,0)=d(x,0)+d(y,0) implies y=tx for some t>0.
Strict convexity and pseudo strictly convexity are equivalent in normed linear spaces (see e.g. [12] p. 122 or [71])but not in metric linear spaces[71]. (For strict convexity in normed linear spaces one may refer to [37]).

The following example shows that a P.S.C. metric linear space need not be S.C.
Definition 4.1.3 Let (X,d) be a metric space and M a subset of X. Given a non-empty subset A of X, **a best approximation of A by M** is a function $\phi : A \to M$ such that $d(x, \phi(x)) = d(x, M)$ for x in A.

The following theorem deals with the uniqueness of best approximation:
Theorem 4.1.1 [78] Let (X,d) be a convex metric linear space with pseudo strict convexity and let M a subset of X, let $\phi : X \to M$ be a continous best approximation of X by M. Then M is a Chebyshev set.
Proof : Since $\phi : X \to M$ is best approximation of X by M, $d(x, \phi(x)) = d(x, M)$ for all x $\in X$ i.e. $P_M(x)$ is non-empty for each x $\in X$.

Now we show that $P_M(x)$ is a singleton . For any x in X, let m$\in P_M(x)$. Suppose y is on the line segment [m,x] and u$\in P_M(y)$. Then

$$
\begin{aligned}
d(u,x) &\leq d(u, y) + d(y, x) \\
&\leq d(m, y) + d(y, x) \\
&= d(m,x) \qquad \text{(as y} \in \text{[m,x))} \\
&\leq d(u,x) \text{ as m} \in P_M(x).
\end{aligned}
$$

Therefore , the inequalities are all equalities and so d(x,u)=d(x,m)=d(x,M) i.e. u $\in P_M(x)$ and therefore $P_M(y) \subseteq P_M(x)$.

Also m $\in P_M(y)$ as d(y ,u)+ d(y ,x)= d(m ,y)+ d(y ,x) implies d(y ,u)= d(y ,m).

Since d(u,x)=d(u,y)+d(y,x) i.e. d(u-x,0) = d(u-y,0)+ d(y-x,0), a consequence of pseudo strict convexity is that u, y and x are collinear. (By P.S.C. u-y= t(y-x) i.e. $y = u/(1+t) + tx/(1+t)$ and therefore u=m as u ,y , x are collinear and d(y,u)=d(y,m) . Hence $P_M(y) = \{m\}$. Since ϕ is directionally continuous at x and $\phi([m,x]) = \{m\}$, it follows that $\phi(x) = m$. Thus $P_M(y) = \{\phi(x)\}$ is a singleton set i.e M is a Chebyshev set.

Note : The following result is established in the proof of Theorem 4.1.1 without the requirement of pseudo strict convexity.

Let (X,d) be a convex metric linear space , M a subset of X,x an element of X and m an element of $P_M(x)$. Then for each y$\in$ [m,x) , $\{m\} \subseteq P_M(y) \subseteq P_M(x)$.

Definition 4.2.1 Let (X,d) be a metric space , M a subset of X. The mapping $P_M : X \to M$ defined by

$$P_M(x) = \{m \in M : d(x,m) = d(x,M)\}$$

Is called **the multi – valued metric projection** of X onto M.

By U_M , we denote the set

$$U_M = \{ x \in X: \text{card } P_M(x) \leq 1 \} .$$

If card $P_M(x) \geq 2$, we say that the metric projection is **totally multi-valued** and if $2 \leq$ card $P_M(x) < \infty$, then the metric projection is called **finitely multi-valued** .In the special case when card $P_M(x) = \chi_0$, we say that the metric projection is **countably multi-valued.**

Definition 4.2.2 If P_M is a totally multivalued metric project then the set M is **strongly proximinal .**

It is clear that every strong proximinal is proximinal and hence closed. (Let M be not closed. Let $x \in \overline{M}/M$. Then d(x , M)=0. Since M is proximinal , there exists m$\in$M such that d(x, m) =d(x, M)=0 and so x=m i.e. x $\in$M , a contradiction).

It was shown in [49] that if (X,d) Is a strictly convex metric space, then the corresponding P_M a single valued metric projection. A question that arises is what happens in spaces which are not strictly convex. We have:

Theorem 4.2.1 [78] Let M be an arbitrary non-void proper subset of a convex metric space (X, d). Then P_M is not a finitely multivalued metric projection.

Proof Case 1 If $\overline{M} = X$ then for every $x \in X\backslash M$, we have $P_M(x) = \phi$ and hence P_M is not a finitely multivalued metric projection.

Case 11 If $\overline{M} \neq X$, let $x_0 \in X\backslash\overline{M}$ then there exists a neighbourhood

of x_0 contained in $X\backslash\overline{M}$.

Let $r = d(x_0, M) > 0$. Suppose P_M is a finitely multivalued metric projection.

Let $P_M(x_0) = \{m_1, m_2, \ldots, m_k, k \geq 2\}$.

Then $d(x_0, m_i) = r$, $I = 1, 2, \ldots, k$.

Let $y_0 = W(x_0, m_i, \lambda)$, $0 < \lambda < 1$ and $0 < 3\lambda r < \min_{2 \leq i \leq k} d(m_1, m_i)$.

We claim that

(1) $B(y_0, \lambda r) \subseteq B(x_0, r)$

(2) $B(y_0, \lambda r) \cap M = \{m_1\}$

(1)   Let $x \in B(y_0, \lambda r)$ then

$$d(x, x_0) \leq d(x, y_0) + d(y_0, x_0)$$
$$= d(x, y_0) + d(W(x_0, m_1, \lambda), x_0)$$
$$\leq d(x, y_0) + \lambda d(x_0, x_0) + (1 - \lambda) d(m_1, x_0)$$
$$\leq \lambda r + \lambda.0 + (1 - \lambda)r$$
$$= r.$$

This implies that $x \in B(x_0, r)$. Hence $B(y_0, \lambda r) \subseteq B(x_0, r)$.

(2)   Since $B(y_0, \lambda r) \subseteq B(x_0, r)$ and $B(x_0, r) \cap M = \{m_{1,} m_2, \ldots, m_k\}$, it follows that

$B(y_0, \lambda r) \cap M = \{m_1, m_2, \ldots, m_k\}$.

We first show that

(a) $m_1 \in B(y_0, \lambda r)$

(b) $m_i \notin B(y_0, \lambda r), i = 1, 2, ..., k$

(a)   Consider
$$d(y_0, m_1) = d(W(x_0, m_1, \lambda), m_1)$$
$$\leq \lambda d(x_0, m_1) + (1 - \lambda) d(m_1, m_1)$$
$$= \lambda r$$
Which implies that $m_1 \in B(y_0, \lambda r)$

(b)   Since
$$d(m_1, m_i) \leq d(m_1, y_0) + d(y_0, m_i)$$
$$d(y_{0,} m_i) \geq d(m_1, m_i) - d(m_1, y_0)$$
$$\geq 3\lambda r - \lambda r$$
$$= 2\lambda r$$
$$> \lambda r$$

So $m_i \notin B(y_0, \lambda r), i = 1, 2, ..., k$

From (a) and (b), it follows that $B(y_0, \lambda r) \cap M = \{m_1\}$.

We claim that $y_0 \in U_M$. Consider

$$d(y_0, m_1) = d(W(x_0, m_1, \lambda), m_1)$$

$$\leq \lambda d(x_0, m_1) + (1 - \lambda) d(m_1, m_1)$$

$$= \lambda r.$$

Thus $d(y_0, m_1) \leq \lambda r$ and so $d(y_0, M) \leq \lambda r$ (4.2.1)

Now   let $d(y_0, M) \leq \lambda r$. Then $d(y_0, m) = d(y_0, M) \leq \lambda r$ (using   (4.2.1))
$\Rightarrow m \in B(y_0, \lambda r) \Rightarrow \Rightarrow m \in B(y_0, \lambda r) \cap M = \{m_1\}$. So $P_M(y_0) \subseteq \{m_1\}$ and hence $card P_M(y_0) \leq 1$
which implies that $y_0 \in U_M$. Thus P_M is not a finitely multi-valued metric projrction
contradicting the hypothesis.

55

<u>**Remark 4.1.1**</u> For normed linear spaces this result was proved in [72]:

It may be remarked that for a convex set M in a convex metric space the corresponding P_M isn't a finitely multi-valued metric projection.

Indeed, if M is a convex set, $x \in X \setminus M$ and if $m_1, m_2 \in P_M(x)$ with $m_1 \neq m_2$, then for every $\lambda \in (0,1)$ we have

$$d(x, W(m_1, m_2, \lambda)) \leq \lambda d(x, m_1) + (1 - \lambda) d(x, m_2)$$

$$= \lambda d(x, M) + (1 - \lambda) d(x, M)$$

$$= d(x, M)$$

Also by the definition of $d(x, M)$ we have $d(x, M) \leq d(x, W(m_1, m_2, \lambda))$ as $W(m_1, m_2, \lambda) \in M$ by the convexity of M.

Hence $d(x, W(m_1, m_2, \lambda)) = d(x, M)$ i.e. $W(m_1, m_2, \lambda) \in M$ for every $\lambda \in (0,1)$.

The. following result on metric projection was proved by Ioan Serb [72].

Let M be a non-void proper subset of a metrizable vector space X. If

P_M is a countable multi-valued metric projection, then P_M is a perfect subset of X. Using this we have:

<u>**Theorem 4.2.2**</u> [81] If (X,d) is a complete metric linear space and M a non-void proper subset of a X then P_M cannot be a countably multi-valued metric projection.

Proof We suppose that there exists a set $M \subseteq X$ with the property that P_M is a countably multi-valued metric projection. So by the above result of Ioan Serb, M is a perfect set. If $x \in X \setminus M$ then $cardP_M(x) = \chi_0$. We claim that $P_M(x) = \varepsilon$, where $S(x, d(x, M))$ is the sphere with centre x and radius d(x , M) , is a perfect set.

(a) $P_M(x)$ is a closed set as it is an intersection of two closed sets.

(b) $P_M(x)$ is dense in itself.

Indeed, if m_0 is an isolated point of $P_M(x)$, then there exists a ball $B(m_0, \varepsilon)$ with centre m_0 and radius $\varepsilon > 0$ such that $P_M(x) \cap B(m_0, \varepsilon) = \{m_0\}$.

Let us consider the point $x_\lambda = W(x, m_0, \lambda)$ with $0 < \lambda < \varepsilon \backslash [2d(x, m_0)] < 1$ since

$$d(x, x_\lambda) = d(x, W(x, m_0, \lambda))$$

$$\leq \lambda d(x, x) + (1 - \lambda) d(x, m_0)$$

$$= (1 - \lambda) d(x, m_0)$$

$$= (1 - \lambda) d(x, M) \qquad (4.2.2)$$

It follows that $x_\lambda \in X \backslash M$.

Now

$$d(x_\lambda, m_0) = d(W(x, m_0, \lambda), m_0)$$

$$= \lambda d(x, m_0) + (1 - \lambda) d(m_0, m_0)$$

$$= \lambda d(x, m_0) \qquad (4.2.3)$$

$$< \varepsilon \backslash 2.$$

On the other hand, let $m \in M, m \neq m_0$. Then $m \in P_M(x) \backslash \{m_0\}$ or $m \notin B(x, d(x, M))$. We shall prove that $d(x, m_0) > d(x_\lambda, m_0)$ in both the cases.

If $m \in P_M(x) \backslash \{m_0\}$, we have $d(x_\lambda, m) \geq |d(m_0, m) - d(x_\lambda, m_0)|$.

The proof of which is as under :

$$d(x_\lambda, m_0) = d((x_\lambda - m) + (m - m_0), 0)$$

$$\leq d((x_\lambda - m), 0) + d((m - m_0), 0)$$

$$\Rightarrow d(x_\lambda, m) \geq d(x_\lambda, m_0) - d(m, m_0) \qquad (4.2.4)$$

and

$$d(m_0, m) = d((m_0 - x_\lambda) + (x_\lambda - m), 0)$$

$$\leq d(m_0, x_\lambda) + d(x_\lambda, m)$$

$$\Rightarrow d(x_\lambda, m) \geq d(m, m_0) - d(m_0, x_\lambda) \qquad (4.2.5)$$

Combining $(4.2.4)$ and $(4.2.5)$, we get the result.

So $d(x_\lambda, m) \geq |d(m_0, m) - d(x_\lambda, m_0)|$

$$> \varepsilon - \varepsilon/2 \, (\text{Since } m \notin \{m_0\} = P_M(x) \cap B(m_0, \varepsilon) \text{ and } m \in P_M(x))$$

$$= \varepsilon/2$$

$$> d(x_\lambda, m_0).$$

If $m \notin B(x, d(x, M))$ we have

$$d(x_\lambda, m) \geq |d(x, m) - d(x, x_\lambda)|$$

$$> \lambda d(x, M) - (1 - \lambda) d(x, M) \, (\text{as } (4.2.2) \Rightarrow d(x, x_\lambda) \leq (1 - \lambda) d(x, M))$$

$$= \lambda d(x, M)$$

$$\geq d(x_\lambda, m_0) \, (\text{as } (4.2.3) \Rightarrow d(x_\lambda, m_0) \leq d(x, m_0) = \lambda d(x, M)).$$

Hence in both the cases we get that $d(x_\lambda, m) > d(x_\lambda, m_0)$. It follows that $P_M(x_\lambda) = \{m_0\}$ and $x_\lambda \in X \setminus M$. So P_M is not countably multi-valued, a contradiction. Therefore $P_M(x)$ has no isolated point. Thus $P_M(x)$ is dense in itself and hence a perfect set. But if $P_M(x)$ is a perfect set of a complete metric space X, then $P_M(x)$ is an uncountable set (see [30],p.72), contradicting our supposition. The theorem is thus proved.

Remark 4.2.2 For Banach space this result was proved by Ioan Serb[72].

Next we shall give a characterization of the semi-metric linear spaces which aren't metric linear spaces in terms of finitely multivalued metric projections.

Theorem 4.2.3 [81] . In every semi-metric linear space X, which isn't a metric linear space, there exist sets $M_2, M_3, ..., M_n, ...$ as well as the sets A and B such that

$\left(\stackrel{.}{i}\right)$ $cardP_{M_n}(x) = n$, for every $x \in X \setminus M_n$ and every $n \in N$, and

$\left(\stackrel{..}{ii}\right)$ $cardP_A(x) = \chi_0$ for every $x \in X \setminus A$,

$\left(\stackrel{...}{iii}\right)$ $cardP_B(x) = c$ for every $x \in X \setminus B$.

Proof . Since X is a semi-metric but not a metric linear space, there exists an element $x_0 \neq 0$ with $d(x_0, 0) = 0$. We shall prove that the sets

$\quad M_2 = \{x_0, x_0 \setminus 2\},$

$\quad M_3 = \{x_0, x_0 \setminus 2, x_0 \setminus 3\},$

$\quad M_n = \{x_0, x_0 \setminus 2, x_0 \setminus 3, ..., x_0 \setminus n\},$

$\quad A = \{x_0, x_0 \setminus 2, x_0 \setminus 3, ..., x_0 \setminus n, ...\}$

and $B = \{\lambda x_0\}_{\lambda \in (0,1)}$ have the properties $\left(\stackrel{.}{i}\right)$ ---- $\left(\stackrel{...}{iii}\right)$ respectively.

$\left(\stackrel{.}{i}\right)$ To prove $CardP_{M_n}(x) = n$, for every $x \in X \setminus M_n$ and every $n \in N$.

Let $x \in X \setminus M_n$, then we have

d(x , m) = d(x-m ,0)

$\qquad \leq$ d(x ,0)+ d(-m ,0)

$\qquad =$ d(x ,0)+ d(m ,0)   (4.2.6)

We claim that d(m,0) = 0 for all $m \in M_n$.

$m \in M_n \Rightarrow m = x_0 \setminus n$ for same natural number n.

Now for $\lambda \in (0,1)$, We have

$$d(\lambda x_0, 0) = d(\lambda x_0 + (1-\lambda)0, 0)$$
$$\leq \lambda d(x_0, 0) + (1-\lambda)d(0,0)$$
$$= \lambda d(x_0, 0).$$

For $\lambda = 1 \backslash n$, We get $d((1 \backslash n)x_0, 0) \leq (1 \backslash n)d(x_0, 0) = 0$.

Also $d(x_0 \backslash n, 0) \geq 0$ implies that $d(x_0 \backslash n, 0) = 0$ i.e. d(m,0)=0 for all $m \in M_n$. So , (4.2.6) implies $d(x, m) \leq d(x, 0)$.

Also $d(x, 0) = d(x - m + m, 0)$
$$\leq d(x - m, 0) + d(m, 0)$$
$$= d(x, m).$$

Hence, we have $d(x, 0) = d(x, m)$ for every $m \in M_n$ and so $cardP_{M_n}(x) = cardM_n = n$.

$\left(\overset{\cdot\cdot}{\text{ii}}\right)$ As discussed above for any $m \in A$, we have $m = x_0 \backslash n, n \in N$ and $d(x, 0) = d(x, m)$ for all $m \in A$ which, further implies that $cardP_A(x) = cardA$.

Since A is infinite countable set, by definition of χ_0, we have $cardP_A(x) = cardA = \chi_0$.

$\left(\overset{\cdot\cdot\cdot}{\text{iii}}\right)$ For $m \in B$ we have $m = \lambda x_0$ for $0 < \lambda < 1$. Again as discussed in $\left(\overset{\cdot}{\text{i}}\right)$ we have

$d(x, 0) = d(x, m)$ for all $m \in B$ which implies $cardP_B(x) = B$. Since B is an infinite uncountable set, $cardP_B(x) = cardB$.

Corollary 4.2.1. The convex semi-metric linear space (X,d) isn't a metric linear space if and only if there exists a subset M of X such that P_M is a finitely multi-valued metric projection.

Proof Since X is a convex semi-metric linear space but not a metric linear space, there exists an element $x_0 \neq 0$ with $d(x_0, 0) = 0$. Then as discussed in Theorem 4.2.3, there exists a set $M = M_n = \{x_0, x_0 \backslash 2, x_0 \backslash 3, ..., x_0 \backslash n\}$ with the property that $cardP_{M_n}(x) = cardM_n = n$ i.e. P_M is a finitely multi-valued metric projection.

Conversly, suppose that there exists a subset M of X such that P_M is a finitely multi-valued metric projection. We are to prove that X isn't a metric linear space.

On contrary, if we assume X to be a metric linear space, then we get from Theorem 4.2.1 that P_M isn't a finitely multi-valued metric projection, a contradiction to the hypothesis.

Remark 4.2.3 For semi-normed spaces the above result was proved in [72].

Considering M to be strongly proximinal subset of a Banach space X, Ioan Serb [73] proved that $cardP_B(x) \geq c$ for every $x \in X \setminus M$, and the completeness of the space is essential for the validity of the result. In [74], the same result was proved for complete metrizable locally convex spaces i.e. for Frachet spaces.

In convex metric spaces we have:

Theorem 4.2.4 [81]. If M is strongly proximinal set in a complete convex metric space (X, d), then $cardP_B(x) \geq c$ for all $x \in X \setminus M$,

Proof. Since M is strongly proximinal set, M is closed and so is $P_M(x) = M \cap B(x, d(x, M))$. We shall show that if $x \in X \setminus M$, $P_M(x)$ does not contain isolated points.

Suppose $m_0 \in P_M(x)$ is an isolated point of $P_M(x)$ for a given $x \in X \setminus M$. Then there exists an $\varepsilon \in (0.1)$ such that $B(m_0, \varepsilon d(x, M)) \cap P_M(x) = \{m_0\}$. (4.2.7)

Let $x_0 = W(x, m_0, \varepsilon \setminus 3)$, we have

$$d(x, x_0) = d(x, W(x, m_0, \varepsilon \setminus 3))$$
$$\leq (1 - \varepsilon \setminus 3) d(x, m_0) \qquad \text{(by the convexity of X)}$$
$$= (1 - \varepsilon \setminus 3) d(x, M)$$
$$< d(x, M),$$

It follows that $x_0 \in X \setminus M$. On the other hand

$$d(x_0, m_0) = d(W(x, m_0, \varepsilon \setminus 3), m_0)$$
$$\leq \varepsilon \setminus 3 d(x, m_0) \qquad \text{(by the convexity of X)}$$
$$= (\varepsilon \setminus 3) d(x, M). \qquad (4.2.8)$$

This implies

$$d(x_0, m) \leq (\varepsilon \setminus 3) d(x, M). \qquad (4.2.9)$$

Let $m \in M$. If $m \notin P_M(x)$ we have

$$d(x_0, m) \geq d(x, m) - d(x_0, x)$$

$$> d(x,M) - d(x,x_0)$$
$$\geq d(x,M) - (1 - \varepsilon \backslash 3)d(x,M)$$
$$= (\varepsilon \backslash 3)d(x,M)$$

i.e. $\qquad\qquad d(x_0,m) > (\varepsilon \backslash 3)d(x,M)$,

so $m \notin P_M(x_0)$ (if $m \in P_M(x_0)$ then $d(x_0,m) \geq d(m_0,m) - d(m_0,x_0)$ $\leq (\varepsilon \backslash 3)d(x,M)$ by (4.2.9))

If $m \in P_M(x)\backslash\{m_0\}$, we have

$$d(x_0,m) \geq d(m_0,m) - d(m_0,x)$$
$$> \varepsilon d(x,M) - (\varepsilon \backslash 3)d(x,M) \qquad\qquad \text{(by (4.2.7) and (4.2.8))}$$
$$= 2(\varepsilon \backslash 3)d(x,M)$$

and so $\quad m \notin P_M(x_0)$.

Thus for all $m \neq m_0, m \in M$, we have $m \notin P_M(x_0)$ and it follows that $P_M(x) = \{m_0\}$ and this contradicts the fact that M is a strongly proximinal set.

Thus $P_M(x)$ is a closed set and has no isolated point i.e. $P_M(x)$ is a perfect set in X for all $x \in X \backslash M$ .Since every perfect subset of a complete metric space has the cardinality at least c (Theorem 6.65 , p.72 [30]), $card P_B(x) \geq c$ for all $x \in X \backslash M$,

--- X ---

CHAPTER – V

ON ε-BIRKHOFF ORTHOGONALITY

AND

ε-NEAR BEST APPROXIMATION

An element x of a normed linear space E is said to be orthogonal (in the sense of Birkhoff) to an element $y \in E$ if $\|x + \alpha y\| \geq \|x\|$ for every scalar α. The notion of Birkhoff orthogonality was used to prove some results on best approximation in normed linear spaces (see e.g. [84]). This notion of orthogonality was extended to metric linear spaces by T.D. Narnag [47] and some results on best approximation were proved. A generalization of Birkhoff Orthogonality, called ε-birkhoff orthogonality was introduced by Sever Silvestru Dragomir [23] in normed linear spaces ($\|x + \alpha y\| \geq (1 - \varepsilon)\|x\|$ for all scalars α) and this notion was used to prove a decomposition theorem ([23]-Theorem 3). We extend the decomposition theorem in metric linear spaces (Theorem 5.1)

For a subset A of a normed linear space X and a +ve number ε, an ε-near best approximation of A by M is a map $\phi : A \to M$ such that $\|x - \phi(x)\| \leq d(x, M) + \varepsilon$ for all x in A. This notion of ε-near best approximation was used by Paul C. Kainen et al [38] to show that the existence of a continuous ε-near best approximation in a strictly convex normed linear space X and taking values in a suitable subset M implies that M has the unique best approximation property. We extend this result to convex metric linear spaces (Theorem 5.2). We also extend some results on ε-near best approximation proved in [23] to metric linear spaces (Theorem 5.3 and its corollaries).

To start with, we recall a few definitions.

Definition 5.1 [53]. Given a non-empty subset A of a metric space (X,d) and a positive number ε, **near best approximation of A by M** is a map ø:A→M such that

d (x,ø(x)) $\leq$ d(x,M) $+ \varepsilon$ for all x in A.

Definition 5.2 [23]. For a normed linear space X, over a field K (K=R or C) and $\varepsilon \in]0,1[$, an element XxX is said to be ε-Birkhoff Orthogoanl to $y \in X$ if

$$\|x + \alpha y\| \geq (1-\varepsilon)\|x\| \text{ for all scalars } \alpha \in K$$

We denote it by $x \perp y (\varepsilon\text{-B})$.

For a metric linear space (X,d) over field K and $\varepsilon \in]0,1[$, an element $x \in X$ is said to be $\varepsilon-$**Birkhoff Orthogonal** to $y \in X$ [23] if

$$d(x + \alpha y, 0) \geq (1-\varepsilon) \, d(x, 0) \text{ for all } \alpha \in K \text{ and we denote it by}$$

$x \perp y (\varepsilon\text{-B})$.

If A is a non-empty subset of X then by $\varepsilon-$**Birkhoff Orthogonal Complement** $A^{\perp}(\varepsilon\text{-B})$, we denote the set of all elements of X which are $\varepsilon-$Birkhoff Orthogonal to A i.e.

$A^{\perp}(\varepsilon\text{-B}) = \{ y \in X : y \perp x(\varepsilon\text{-B}) \text{ for all } x \in A\}$.

Since $A^{\perp}(\varepsilon\text{-B}) = \{ y \in X : y \perp x(\varepsilon\text{-B}) \text{ for all } x \in A\}$, $0 \in A^{\perp}(\varepsilon\text{-B})$ as $0 \perp x (\varepsilon\text{-B})$ for all $x \in A$ (d $(0 + \alpha x, 0) \geq (1-\varepsilon)$ (d $(0,0)$ for all $x \in A$).

We claim that $A \cap A^{\perp}(\varepsilon\text{-B}) \subseteq \{0\}$ for every $\varepsilon \in]0, 1[$.

Let $y \in A \cap A^{\perp}(\varepsilon\text{-B})$. Then $y \cap A^{\perp}(\varepsilon\text{-B})$.

Now $y \cap A^{\perp}(\varepsilon\text{-B}) \Rightarrow y \perp x (\varepsilon\text{-B})$ for all $x \in A$.

$\Rightarrow \quad y \perp y (\varepsilon\text{-B})$

$\Rightarrow \quad d(y + \alpha y) \geq (1-\varepsilon) \, d(y, 0)$ for all $\alpha \in K$

$\Rightarrow \quad 0 \geq (1-\varepsilon) \, d(y, 0)$ by taking $\alpha = -1$

$\Rightarrow \quad d(y, 0) \leq 0 (\text{as } (1-\varepsilon) \geq 0)$

$\Rightarrow \quad d(y, 0) = 0$

$\Rightarrow \quad y = 0$

and so $A \cap A^{\perp}(\varepsilon\text{-B}) \subseteq \{0\}$.

we now prove a lemma needed in the proof of decomposition theorem.

Lemma 5.1 [75]. Let G be a closed linear subspace of a metric linear space (X,d), $G \neq X$. Then for any $\varepsilon \in \ ] \, 0, \, 1 \, [$, the ε–Birkhoff Orthogonal complement of G is non-zero.

Proof. Let $y \in X \backslash G$. Since G is closed, $d\,(y, G) = r > 0$. Thus there exists $y_\varepsilon \in G$ such that

$$r \leq d\,(y, y_\varepsilon) \leq \quad r\,/\,(1\text{-}\varepsilon) \quad \text{(as } r\text{=}d\,(y,G)\,)$$

i.e. $r \leq d\,(y\text{-}y_\varepsilon, \, 0) \leq \quad r\,/\,(1\text{-}\varepsilon)$.

Put $x_\varepsilon = y\text{-}y_\varepsilon$, we have $x_\varepsilon \neq 0$ and for all $y_1 \in G$ and $\lambda \in K$, we obtain

$$
\begin{aligned}
d(x_\varepsilon + \lambda \, y_1, \, 0) \quad &= \quad d\,(y\text{-}y_\varepsilon + \lambda \, y_1, \, 0) \\
&= \quad d\,(y, \ y_\varepsilon \text{-} \lambda y_1) \\
&\geq \quad r \text{ (as } d\,(y,G) = r \text{ and } y_\varepsilon \text{-} \lambda y_1 \in G) \\
&\geq \quad (1\text{-}\varepsilon)\, d\,(x_\varepsilon, \, 0)
\end{aligned}
$$

i.e. $x_\varepsilon \perp y_1 (\varepsilon\text{-}B)$ and so $x_\varepsilon \in G^{\perp}(\varepsilon\text{-}B)$.

Using this lemma, we prove the following decomposition theorem in metric linear spaces (which for normed linear spaces was proved in [23]) :

Theorem 5.1 [75]. Let G be a closed linear subspace of a metric linear space (X,d). Then for any $\varepsilon \in] \, 0,1 \, [$, we have $X = G \oplus G^{\perp}(\varepsilon - B)$

Proof. Suppose $G \neq X$ and $x \in X$. If $x \in G$, then $x = x + 0 \in G + G^{\perp}(\varepsilon - B)$.

If $x \notin G$, then there exists an element $y_\varepsilon \in G$ such that

$$0 < r = d\,(x, G) \leq d\,(x, \ y_\varepsilon) \quad \leq r/(1\text{-}\varepsilon).$$

Since $x_\varepsilon = x \text{-} y_\varepsilon \in G^{\perp}(\varepsilon - B)$ (by the above lemma), we have

$x = y_\varepsilon + x_\varepsilon \in G + G^{\perp}(\varepsilon - B)$.

Since $\{0\} \subseteq G \cap G^{\perp}(\varepsilon - B) \subseteq \{0\}$, we get $X = G \oplus G^{\perp}(\varepsilon - B)$.

The following theorem shows that the continuity of ε-near best approximation is enough to guarantee the uniqueness of best approximation in convex in convex metric linear spaces which are pseudo strictly convex.

Theorem 5.2 [75]. Let (X, d) be a convex metric linear space which is pseudo strictly convex and M a boundedly compact closed subset of X. Suppose that for each $\varepsilon > 0$, there exists a contnuous ε-near best approximation $\phi : X \to M$ of X by M then M is a Chebyshev set.

Proof. Since a boundedly compact closed set in a metric space is proximinal (see [88], p. 283), $P_M(x)$ is non-empty for each $x \in X$. Let $m \in P_M(x)$

We choose a point $x_0 \in X$ with $r = d(x_0, M) > 0$. Given a +ve integer $n \geq 1$, let ϕ_n :$X \to M$ be continuous with

$$d(x, \phi_n(x)) \leq d(x, M) + 1/n \text{ for all x in X.}$$

Then $\phi_n : B(x_0, r) \to M$ and

$d(\phi_n(x), x_0) \geq r$ for all x in the closed ball B (x_0, r).

Let π be a mapping defined by

$\pi(x) = x_0 + r(x - x_0)/d(x, x_0), x \in X.$

We claim that

$\pi = \{ x : d(x, x_0) \geq r \} \to \{ x : d(x, x_0) = r \} \equiv \partial B(x_0, r)$

is a redial retraction i.e.

(i) $d(\pi(x), x_0) = r$

(ii) for x $\in \partial B(x_0, r)$, $\pi(x) = x$.

Consider

$$
\begin{aligned}
d(\pi(x), x_0) &= d(x_0 + r(x - x_0)/d(x, x_0), x_0) \\
&= d(r(x - x_0)/d(x, x_0), 0), \\
&\geq r d(x - x_0, 0)/d(x, x_0), \text{ by the convexity of } (X, d) \\
&= rd(x, x_0)/d(x, x_0) \\
&= r.
\end{aligned}
$$

Thus,

$$d(\pi(x), x_0), \leq r \qquad\qquad (5.1)$$

As $\pi(x) = x_0 + [r(x - x_0)]/d(x - x_0)$

$$= \quad r\,x/d\,(x,\ x_0) + [(1-r)/\,d\,(x-x_0)]\,x_0$$

i.e. $\pi\,(x) \in [x,\ x_0]$ and so

$$d\,(x,\ \pi\,(x))\,) + d\,(\pi\,(x),\ x_0) = d\,(x,\ x_0) \qquad (5.2)$$

Now

$$
\begin{aligned}
d\,(\pi\,(x),\,x) = \quad & d\,(x_0 + [r\,(x-x_0)\,]\,/d\,(x,\ x_0),\,x) \\[4pt]
= \quad & d\,(r\,(x-x_0)\,/d\,(x,\ x_0),\,x-x_0) \\[4pt]
\leq \quad & [1-r/d(x,\ x_0)\,]\,d\,(0,\,x-x_0)\ , \text{ by the convexity of X} \\[4pt]
= \quad & [1-r/d(x,\ x_0)\,]\,d\,(x,\ x_0) \\[4pt]
= \quad & d\,(x,\ x_0) - r.
\end{aligned}
$$

Hence, $-d\,(\pi\,(x),\,x) \geq r - d(x,\ x_0)$, So (5.2) implies

$$d\,(\pi\,(x),\ x_0) \geq d\,(x,\ x_0) + [r - d(x,\ x_0)\,] = r$$

i.e. $d\,(\pi\,(x),\ x_0) \geq r \qquad (5.3)$

Combining (5.1) and (5.3), we get $d\,(\pi\,(x,\ x_0)) = r$.

For $x \in \partial B\,(x_0,r)$ i.e. $d\,(x,\ x_0) = r$, we get

$$\pi\,(x) = x_0 + r\,(x-x_0)/d\,(x,\ x_0)$$

$$= x$$

i.e. $\pi\,(x) = x$ for all $x \in \partial B\,(x_0,r)$.

Thus $\pi : \{\,x : d\,(x,\ x_0) \geq r\} \ \to \ \{\,x : d\,(x,\ x_0) = r\}$ is a radial retraction and $\pi_0 \phi_n : B(x_0, r) \to \partial B\,(x_0, r)$.

Now $\phi_n(x)$, for x in $B\,(x_0,\ r)$ satisfies.

67

$$d\left(\phi_n(\mathrm{x}),\mathrm{x}_0\right) \le d(\mathrm{x},M)+1/n+d(\mathrm{x},\mathrm{x}_0) \qquad (5.4)$$

$$\le d(\mathrm{x},\mathrm{x}_0)+ d(\mathrm{x}_0, M) + 1/n + d(\mathrm{x},\ \mathrm{x}_0)$$

$$= d(\mathrm{x}_0, M) + 1/n + 2d(\mathrm{x},\ \mathrm{x}_0)$$

$$\le 3r + 1.$$

Hence $\phi_n(\ \mathrm{B}(\mathrm{x}_0,r)\ ) \subseteq M\cap \mathrm{B}(\mathrm{x}_0, 3r+1)$ and $\phi_n(\mathrm{B}(\mathrm{x}_0,r)\ )$ is a bounded subset of M. So cl $(\phi_n(\mathrm{B}(\mathrm{x}_0,r)\ )$ is compact since M is given to be boundedly compact.

Let $P:\mathrm{X} \to \mathrm{X}$ be the reflection through x_0.

i.e. $P(y) = \mathrm{x}_0 + (\mathrm{x}_0-y)$ \qquad\qquad (5.5)

Then cl $(P_0\pi_0\phi_n(\mathrm{B}(\mathrm{x}_n(\mathrm{B}(\mathrm{x}_0,r)\ )) = P_0\pi(\mathrm{cl}\phi_n(\mathrm{B}(\mathrm{x}_0,r))$ is compact subset of $\partial \mathrm{B}[\mathrm{x}_0,r]$ and $P_0\pi_0\phi_n$ is a continuous function from $\mathrm{B}(\mathrm{x}_0,r)$ into $\partial \mathrm{B}(\mathrm{x}_0,r)$.

Since in a convex metric linear space $\mathrm{B}(\mathrm{x}_0,r)$ is convex, by Rothe's theorem, a version of Schauder's theorem (see [88], p. 27) for each n, $P_0\pi_0\phi_n$ has a fixed point x_n in $\mathrm{B}(\mathrm{x}_0,r)$.

Thus $x_n \quad = \quad P_0\pi_0\phi_n(x_n)$

$$= \quad P_0(\pi_0\phi_n(x_n)\ )$$

$$= \quad 2\mathrm{x}_0 - P_0\pi_0\phi_n(\pi_0\phi_n(x_n)\ )\ (\text{using }(5.5)\ )$$

and so $(\pi_0\phi_n)(x_n) = 2\mathrm{x}_0 - x_n$

We claim that $x_n,\ \mathrm{x}_0,\ 2\ \mathrm{x}_0 - x_n = \pi_0\phi_n(x_n)$ and $\phi_n(x_n)$ are consecutive collinear points.

Since $2\ \mathrm{x}_0 - x_n = \pi_0\phi_n(x_n)$ implies $2\ \mathrm{x}_0 - x_n - \pi_0\phi_n(x_n) = 0$ i.e.

$\alpha\mathrm{x}_0 + \beta x_n + \gamma\pi_0\phi_n(x_n)=0$ with $\alpha+\beta+\gamma = 0$ i.e. $\mathrm{x}_0 + \beta x_n + \gamma\cdot\pi_0\phi_n(x_n)/(\beta+\gamma)$.

Also, by the definition of $\pi(x)$ we have

$$\pi(\phi_n(x_n)) = \mathrm{x}_0 +(r(\phi_n(x_n)-\mathrm{x}_0))/\ d(\phi_n(x_n),\mathrm{x}_0)$$

$$= r\phi_n(x_n)/d(\phi_n(x_n),\mathrm{x}_0)+(1-r/[d(\phi_n(x_n),\mathrm{x}_0]\mathrm{x}_0)$$

$$\Rightarrow 1.\pi_0\phi_n(x_n)-r\phi_n(x_n)/d(\phi_n(x_n),\mathrm{x}_0)-(1-r/d(\phi_n(x_n),\mathrm{x}_0))\mathrm{x}_0 =0$$

$$\Rightarrow \alpha.\pi_0\phi_n(x_n)+\beta\phi_n(x_n)+\gamma.\mathrm{x}_0 =0$$

with $\alpha + \beta + \gamma = 1 - r / d(\phi_n(x_n), x_0) - 1 + r / d(\phi_n(x_n), x_0) = 0$

$\Rightarrow \pi(\phi_n(x_n)) = (\beta\phi_n(x_n) + \gamma.x_0) / (\beta + \gamma)$

and so

$d(\phi_n(x_n), x_n) \geq d(\pi_0\phi_n(x_n), x_n)$

$$= d(2x_0 - x_n, x_n)$$

$$= d(x_n, x_0) + d(x_0, 2x_0 - x_n),\ as\ x_n, x_0$$

$$\text{and } 2x_0 - x_n \text{ are collinear}$$

$$= d(x_n, x_0) + d(x_n, x_0)$$

$$= 2d(x_n, x_0)$$

Now we prove that $d(x_n, x_0) = r$

Since $\pi_0\phi_n : B(x_0, r) \to \partial B(x_0, r)\ and\ x_n \in B(x_0, r)$

implies $(\pi_0\phi_n(x_n)x_0) = r\ i.e\ d(2x_0 - x_n, x_0) = r$, i.e. $d(x_n, x_0) = r$.

Hence $d(\phi_n(x_n), x_n) \geq 2r$.

In addition for each m in M,

$d(x_n, m) \geq d(x_n, \phi_n(x_n)) - 1/n$ (using (3.4))

$$\geq 2r - 1/n \qquad\qquad\qquad (5.6)$$

Again M is boundedly compact, the sequence $\{\phi_n(x_n)\}$ in $M \cap B(x_0, 3r+1)$ has a convergent subsequent with limit u in x. Then the sequence $\{P_o\pi_0\phi_n(x_n)\}$ has a convergent subsequence with limit $P_o\pi(u) = x_\infty \in \partial B(x_0, r)$.

Moreover, for each m in M,

$$d((x_\infty - x_0) + (x_0 - m), 0) = d((x_\infty - m, 0)$$

$$= d(x_\infty, m)$$

$$\geq 2r(\text{using } (5.6)) \qquad\qquad (5.7)$$

69

If m is in $P_M(x_o)$ then $d(x_0,m) = d(x_0,m) = r$.

Also $d(x_\infty, x_0) = r$ as $x_\infty \in \partial B(x_0, r)$. So

$$d((x_\infty - x_0) + (x_0 - m),0) = d(x_\infty - x_0, m - x_0)$$

$$\leq d(x_\infty - x_0, 0) + d(m - x_0, 0)$$

$$= r + r$$

$$= 2r$$

Implies

$$d((x_\infty - x_0) + d(x_0 - m),0) \leq 2r \qquad (5.8)$$

Combining (5.7) and (5.8) we have

$$d((x_\infty - x_0) + d(x_0 - m),0) = 2r$$

$$= r + r$$

$$= d((x_\infty - x_0),0) + d((x_0 - m),0) \quad (5.9)$$

Since (X, d) is pseudo strictly Convex, (5.9) implies

$$x_\infty - x_0 = t(x_0 - m) \text{ for some } t > 0.$$

i.e. m = $[(1+t) x_0 - x_\infty]/t$ implying $P_M(x_0) = [(1+t) x_0 - x_\infty]/t$ for $t > 0$. Hence M is Chebyshev.

In strictly convex normed linear spaces this theorem was proved by Paul C. Kainen et al [38] and the above proof is an extension of the one given in [38].

Corollary 5.1 [75]. Let (X, d) be a convex metric linear space, M a boundedly compact subset of X and x an element of X with $r = d(x_0, M) > 0$. Suppose that for some ε, with $0 < \varepsilon < 2r$ there exists continuous ε-near best approximation $\phi : B(x,r) \to M$ of B (x, r) by M. Then there exists a point x_1 in $\partial B(x,r)$ such that d $(x_1, m) \geq 2r - \varepsilon$.

Proof. The proof of this is contained in the first part of the proof of Theorem 5.2 (upto equation (5.6)).

70

If M is an approximatively compact set in a metric space, then $P_M(x)$ is compact for each x in X. Indeed, any $\{m_n\}$ in $P_M(x)$ is a sequence in M with $d(x, m_n) = d(x, M)$ and by the definition of approximative compactness, has a convergent subsequence with limit in M and hence in $P_M(x)$. Using this we have:

Theorem 5.3 [75]. Let M be an approximatively compact set in a metric linear space (X, d) and x an element of X. Suppose that for each $\varepsilon > 0$, there is a continuous ε-near best approximation $\phi_\varepsilon : (x) \cup P_M(x) \to M$. Then $P_M(x)$ is connected.

For normed linear spaces the proof of Theorem 5.3 is given in [38] and that proof can easily be extended to metric linear spaces.

Corollary 5.2 [75]. Let (X, d) be a metric linear space and M an approximately (i.e.. $P_M(x)$ is non-empty and countable for each x in X). Suppose that for each $\varepsilon > 0$ there exists a continuous ε-near best approximation $\phi : X \to M$ of X by M. Then M is a Chebyshev set.

Proof. By Theorem 5.3 for each x, $P_M(x)$ is connected and since the only countable connected set is a singleton, M is Chebyshev,

Corollary 5.3 [75]. Let (X, d) be a metric linear space, M a closed, boundary compact subset of X, and x an element of X with r = d (x, M) > 0. If for each $\varepsilon > 0$, there exists a continuous ε near best approximation $\phi : B(x,r) \to M$ of B(x ,r) by M then $P_M(x)$ is connected.

Proof Since a closed , boundedly compact subset is approximatively compact ([84],p. 383), proof follows from Theorem 5.3 .

--- X ---

71

CHAPTER-VI

ON ε-SIMULTANEOUS APPROXIMATION AND BEST

SIMULTANEOUS CO-APPROXIMATION

This chapter deals with ε-Simultaneous Approximation and Best Simultaneous Co-Approximation. The chapter has been divided into two sections. In the first section we discuss ε-simultaneous approximation. The problem of best simultaneous opprozined (b.s.a.) is concerned with approximating simultaneously any two elements x_1 , x_2 of a space X by the elements of a subset A of X. More generally, if a set of elements B is given in X, one might like to approximate all the elements of B simultaneously by a single element of A. This type of problem arises when a function being approximated is not known precisely, but is known to belong to a set. C. B. Dunhan [24] seems to have been the first who studied the problem of b.s.a. in normed linear spaces. R.C. Buck [18] studied the problem of ε-approximation which reduces to the problem of best approximation for the particular case when ε=0. In this section, we discuss ε-simultaneous approximation for any two elements x_1, x_2 and for a non-empty bounded subset F of a convex metric space (X, d) with respect to a non-empty subset G of X. Defining ε-simultaneous approximation map $P_{G(\varepsilon)} : X \times X \to 2^G$ by

$P_{G(\varepsilon)}(x_1, x_2) = \{g_0 \in G : d(x_1, g_0) + d(x_2 g_0) \leqslant r + \varepsilon\}$ where r = inf (d (x_1, g) + d (x_2, g) : g $\in$ G} and we prove the upper semi-continuity of the map $P_{G(\varepsilon)}$. We also prove the convexity, boundedness, closeness and sharshapedness of the set $P_{G(\varepsilon)}(x_1, x_2)$ and of $P_{G(\varepsilon)}(F)$ in the first section.

The second section of this chapter deals with Best Simultaneous Co-approximation A new kind of approximation, called best co-approximation was introduced in normed linear spaces by C. Franchetti and M. Furi [26] in 1972. This study was taken up later by T.D. Narang, P.L. Papini, Geetha S. Rao, Ivan Singer, S.P. Singh and few others (see [55], [56]). Generalizing the concept of best approximation, Geetha S. Rao and R. Sarvanan studied the problem of best simultaneous co-approximation in normed liner spaces in [64]. In this section, we study the problem of best simultaneous co-approximation in convex metric linear spaces and convex metric spaces thereby extending some of the results proved in [64] We have also given some properties of the set $S_G(x,y)$ i.e. the set of all best simultaneous co-approximations to x, y in G. We have proved that for a convex metric space (X, d) and G a convex subset of X, the set $S_G(x,y)$ is convex. We have also proved the upper semi continuity of the mapping $S_G:\{(x,y):x,y\in X\}\to 2^G$ in totally complete metric linear spaces (a notion introduced by T.D. Narang [50]).

6.1 ε-Simultaneous Approximation

This section deals with ε-simultaneous approximation in metric spaces. In this section, we discuss ε-simultaneous approximation of any two elements x_1, x_2 and then of a non-empty bounded subset F of a convex metric space (X, d) with respect to a non-empty subset G of X.

To start with, we recall a few definitions.

Definition 6.1.1. Let (X, d) be a metric space and G a non-empty subset of X. An element $g_0 \in G$ is said to-be (i) an element of ε-**approximation** (ε-a.) to $x \in$ X if

$d(x, g_0) \leq d(x, g) + \varepsilon$ for all $g \in$ G and $\varepsilon > 0$

i.e. $d(x, g_0) \leq \inf\{d(x, g) : g \in G\} + \varepsilon$

The set of all ε-approximations to $x \in$ X from G is denoted by $P_{G(\varepsilon)}(x)$.

(ii)   an element of ε-**simultaneous approximation** (ε-s. a.) to $x_1, x_2 \in$ X from

       G if $d(x_1, g_0) + d(x_2, g_0) \leq$ r $+\varepsilon$

       where r= $\inf\{d(x_1, g) + d(x_2, g) : g \in$ G}

The set of all ε-smultaneous approximations to x_1 and x_2 from G will be denoted by $P_{G(\varepsilon)}(x_1, x_2)$.

Defintion 6.1.2. Let (X, d) be a metric space, G a non empty subset of X and F a non-empty bounded subset of X. For x in X, let

    $d_F(x)$ = Sup {d (y, x) : y $\in$ F}

    D (F, G)   = $\inf(d_F(x) : x \in$ G}, and

$$P_{G(\varepsilon)}(F) \quad = \{g_0 \in G : d_F(g_0) \le D(F,G) + \varepsilon\} \text{ for } \varepsilon > 0$$
$$= \{g_0 \in G : \sup_{y \in F} \le \inf_{g \in G} \sup_{y \in F} d(y,g) + \varepsilon\}.$$

An element $g_0 \in P_{G(\varepsilon)}(F)$ is called ε-**simultaneous approximation (ε-s.a.) of F with respect to G.**

One of the advantages of considering the sets $P_{G(\varepsilon)}(x_1, x_2)$ and $P_{G(\varepsilon)}(F)$ with $\varepsilon > 0$, instead of the sets $P_G(x_1, x_2)$ and $P_G(F)$ respectively is that the sets $P_{G(\varepsilon)}(x_1, x_2)$ and $P_{G(\varepsilon)}(F)$ are always non-void for $\varepsilon > 0$.

The problem of ε-s.a: is equivalent to the problem of minimzing certain functional as shown below:

Lemma 6.1.1 [77]. If G is any subset of a metric space (X, d) and F a bounded subset of X. Then the functional $\phi : G \to R$ defined by $\phi(g) = \sup_{f \in F} d(f, g)$ is

continuous.

Proof. Let $\varepsilon > 0$ be given. For any $f \in F$ and g, g' $\in$ G, we have d (f, g) $\le$ d (f, g') +d (g' g) and so $\sup_{f \in F}$ d(f, g) $\le \sup_{f \in F}$ sup d (f, g') + d (g', g)

i.e. $\phi : (g) - \phi (g') \le d (g', g)$

Inerchanging g and g', we get $\phi (g') - \phi (g) \le d (g, g')$ and so $|\phi(g) - \phi(g')| \le d(g, g')$. Therefore, if d (g, g') $< \varepsilon$ then $|\phi(g) - \phi(g')| < \varepsilon$ and consequently ϕ is continuous.

If we take $\phi'(g) = \phi(g) + \varepsilon$ then $\inf_{g \in G} \phi'(g) \inf_{g \in G} \phi g) + \varepsilon$.

So a $g_0 \in G$ satisfying $\phi'(g_0) = \inf_{g \in G} \phi'(g)$ is an ε-s.a. to F

Thus, the problem of ε-s.a. is the problem of minimizing the functional ϕ' on G.

The following lemma deals with the boundedness and closedness of the set $P_{G(\varepsilon)}(F)$.

Lemma 6.1.2. [77] The set $P_{G(\varepsilon)}(F)$ is bounded and is a closed subset of G if G is closed. In addition, $P_{G(\varepsilon)}(F)$, is compact if G is compact.

Proof. Let g_0, $g_0' \in P_{G(\varepsilon)}(F)$. Then

$$d\,(g_0, g_0') \;\leq\; d\,(g_0, y) + d\,(y,\, g_0') \text{ for every } y \in F$$
$$\leq\; d_F\,(g_0) + d_F\,(g_0')$$
$$\leq\; D\,(F,G) + \varepsilon + D\,(F,G) + \varepsilon + D\,(F,D) + \varepsilon \text{ as } g_0,\, g_0' \in P_{G(\varepsilon)}(F)$$
$$=\; 2\,(D\,(F,\,G) + \varepsilon\,).$$

and so $P_{G(\varepsilon)}(F)$ is bounded.

Suppose G is closed. Let g_0 be a limit point of $P_{G(\varepsilon)}(F)$. Then there exists a sequence $<g_0^{\{n\}}>$ in $P_{G(\varepsilon)}(F)$ such that $<g_0^{\{n\}}> \;\to\; g_0$. Now $\Rightarrow g_0^n \in P_{G(\varepsilon)}(F) \Rightarrow$ $d_F(g_0^n) \leq D(F,G) + \varepsilon$ for all $n \Rightarrow \lim\limits_{n\to\infty} d_F(g_0^n) \leq D(F,G) + \varepsilon \Rightarrow d_F(g_0) \leq D(F,G) + \varepsilon$ $\Rightarrow g_0 \in P_{G(\varepsilon)}(F)$ as G being closed. $g_0 \in G$. Therefore $P_{G(\varepsilon)}(F)$ is a closed subset of G. If the set G is compact then the set $P_{G(\varepsilon)}(F)$ is compact as closed subset of a compact set is compact.

If we take $F = (x_1, x_2)$, we have:

Corollary 6.1.1 [77]. The set $P_{G(\varepsilon)}(x_1, x_2)$ is bounded and a closed subset of G if G is closed. In addition, $P_{G(\varepsilon)}(x_1, x_2)$ is compact if G is compact.

The following result shows the convexity of the set $P_{G(\varepsilon)}(F)$ in convex metric spaces.

Propostion 6.1.1. [82]. For any convex set G in a convex metric space (X, d), the set $P_{G(\varepsilon)}(F)$ is convex.

Proof. Let g_0, $g_0' \in P_{G(\varepsilon)}(F)$. Then $d_F(g_0) \leq D(F,G) + \varepsilon$ and $d_F(g_0') \leq D(F,G) + \varepsilon$. For any $f \in F$, consider $d\,(f,\, W(g_0, g_0', \lambda)) \leq \lambda\, d(f, g_0) + (1-\lambda)d\,(f,\, g_0')$. This implies

$$d_F\,(W(g_0, g_0', \lambda)) \;=\; \sup_{f \in F} d\,(f,\, W(g_0, g_0', \lambda))$$

$$\leq \lambda\, \sup_{f \in F} d\,(f,\, g_0) + (1-\lambda)\, \sup_{f \in F} d\,(f,\, g_0')$$

$$= \lambda d_F(g_0) + (1-\lambda)d_F(g_0')$$

$$\leq \lambda\,(D(F,G) + \varepsilon) + (1-\lambda)(D(F,G) + \varepsilon)$$

75

$$= D(F,G)+\varepsilon,$$

Where $W(g_0,g_0',\lambda) \in G$ by the convexity of G, implying that $W(g_0,g_0',\lambda)$ is ε-s.a. in G to F and so the set $P_{G(\varepsilon)}(F)$ is convex.

This proposition shows that if G is a convex subset of a convex metric space (X, d) and if g_0, g_0' are ε-s.a. in G to F then $W(g_0,g_0',\lambda)$ is also ε-s.a. in G to F for every $\lambda \in I$.

For F = $\{x_1, x_2\}$, we get:

Corollary 6.1.2 [76] For any convex set G in a convex metric space (X, d), the set $P_{G(\varepsilon)}(x_1, x_2)$ is convex.

The above corollary shows that in a convex metric space (X, d) if g_0, g_0' are ε-simultaneous approximations to x_1 and x_2 by elements of a convex set, then $W(g_0,g_0',\lambda)$ is also ε-s.a. to x_1 and x_2 for every $\lambda \in I$.

Next result proves the starshapedness of the set $P_{G(\varepsilon)}(F)$

Proposition 6.1.2 [77]. In a convex metric space (X, d), if G is starshaped with respect to g_0 and F a bounded subset of X then $P_{G(\varepsilon)}(F)$ is also starshaped with respect to g_0 provided $g_0 \in P_{G(\varepsilon)}(F)$.

Proof. Let $y \in P_{G(\varepsilon)}(F)$. Then $d_F(y) \le D(F, G) + \varepsilon$. Since G is starshaped with respect to g_0, $W(g_0,g_0',\lambda) \in G$ for $\lambda \in I$.

We claim that $W(g_0,g_0',\lambda) \in P_{G(\varepsilon)}(F)$ for all $\lambda \in I$.

Consider

$$d_F(W(g_0,g_0',\lambda) \quad = \sup_{f \in F} d(f, W(y, g_0, \lambda))$$

$$\le \sup_{f \in F} d(f, y) + (1-\lambda) \sup_{f \in F} d(f, g_0)$$

$$= \lambda\, d_F(y) + (1-\lambda) d_F(g_0)$$

$$\le \lambda\, (D(F,G)+\varepsilon) + (1-\lambda)(D(F,G)+\varepsilon)$$

$$= D(F,G)+\varepsilon$$

Hence

$$d_F W(y, g_0, \lambda) \le D(F,G)+\varepsilon$$

76

implying that $W(y, g_0, \lambda) \in P_{G(\varepsilon)}(F)$ for all $y \in P_{G(\varepsilon)}(F)$ and $\lambda \in I$ i.e. set $P_{G(\varepsilon)}(F)$ is starshaped with respect to g_0.

For F = (x_1, x_2), we get:

Corollary 6.1.3 [76]. In a conved metric space (X, d) if G is starshaped with respect to g_0 then $P_{G(\varepsilon)}(x_1, x_2)$, is also starshaped with respect to g₀ if $g_0 \in P_{G(\varepsilon)}(x_1, x_2)$.

Let CB (X) be the family of non-empty closed and bounded subsets of X. Let H be a **Hausdorff metric** on CB (X) i.e. for A, B∈CB(X),

$$H(A,B) = \text{Max} \left(\sup_{a \in A} d(a,B), \sup_{b \in B} d(b,A) \right)$$

Sastry and Naidu [68], and Govindrajulu [28] proved that under certain conditions, the best simultaneous approximation operator (not necessarily single-valued) is upper semi-continuous. The following result deals with the upper semi-continuity of the ε-simultaneous approximation map:

Proposition 6.1.3 [77]. If G is a compact subset of a metric space (X, d) then the ε-simultaneous approximation map $P_{G(\varepsilon)}$:X→ 2^G is upper semi-continuous i.e. the set K={F∈CB(X) : $P_{G(\varepsilon)}(F) \cap N \neq \phi$} is closed for every closed set N in X.

Proof. Let $\{F_n\}$ be a sequence in K converging to F∈CB (X).

Then there exists a sequence $< x_n >$ in G such that $x_n \in P_{G(\varepsilon)}(F_n) \cap N$ for each n and so $d_F(x_n) \leq D(F_n, G) + \varepsilon$ Consider

$$d_F(x_n) \leq H(F, F_n) + d_F(x_n)$$
$$\leq H(F, F_n) + D(F_n, G) + \varepsilon \qquad (6.1.1)$$

Since G is compact, there exists a subsequence $< x_{n_i} >$ of $< x_n >$ such that $< x_{n_i} > \to x_0$ and so (4.1.1) implies

$$d_F(x_0) \leq D(F, G) + \varepsilon \text{ as } H(F, F_n) \to 0$$

i.e. $x_0 \in P_{G(\varepsilon)}(F)$.

Since$< x_{n_i} > \in$ N and N is closed, $x_0 \in$ N. Consequently

$x_0 \in P_{G(\varepsilon)}(F) \cap$ N i.e. F $\in$ K implying that the map $P_{G(\varepsilon)}$ is upper semi-continuous.

The following result deals with the upper semi-continuity of the ε-simultaneous approximation map $P_{G(\varepsilon)}(x_1, x_2)$:

Proposition 6.1.4 [76]. If G is a compact subset of metric space (X, d), then the ε-simultaneous approximation map

$P_{G(\varepsilon)}$: XxX$\rightarrow 2^G$ is upper semi-continuous i.e. the set B $=\{ (x_1, x_2) \in$ XxX: $P_{G(\varepsilon)}(x_1, x_2) \cap$N $\neq \phi \}$ is closed for every closed set N$\subset$G.

Proof. The proof of Proposition 6.1.4 follows from Proposition 6.1.3 by taking F=(x_1, x_2). However, an independent proof is as under:

Let ($x_1^{(0)}$, $x_2^{(0)}$) be a limit point of the set B. Then there exists a sequence ($x_1^{(n)}$, $x_2^{(n)}$) > in B converging to ($x_1^{(0)}$, $x_2^{(0)}$) $\in$ B, there exists a sequence $<g_n>$ in G such that

$$g_n \in P_{G(\varepsilon)}(x_1^{(n)}, x_2^{(n)}) \cap N, \quad n=1, 2, 3 \ldots$$

Consider

$$
\begin{aligned}
d(x_1^{(0)}, g_n) + d(x_2^{(0)}, g_n) &\leq d(x_1^{(0)}, x_1^{(n)}) + d(x_1^{(n)}, g_n) + d(x_2^{(0)}, x_2^{(n)}) + \\
&\quad d(x_2^{(n)}, g_n) \\
&\leq d(x_1^{(0)}, x_1^{(n)}) + d(x_2^{(0)}, x_2^{(n)}) + \inf \{d(x_1^{(n)}, g) \\
&\quad + d(x_2^{(n)}, g : g \in G\} + \varepsilon. \\
&\leq d(x_1^{(0)}, x_1^{(n)}) + d(x_2^{(0)}, x_2^{(n)}) + \\
&\quad \inf \{d(x_1^{(0)}, x_1^{(n)}) + d(x_1^{(0)}, g) + d(x_2^{(n)}, x_2^{(0)}) + \\
&\quad d(x_2^{(0)}, g) : g \in G\} + \varepsilon \\
&= 2 \{d(x_1^{(0)}, g) + d(x_2^{(0)}, g) : g \in G\} + \varepsilon.
\end{aligned}
$$

This implies

$$\lim [d(x_1^{(0)}, g_n) + d(x_2^{(0)}, g_n)] \leq \inf \{d(x_1^{(0)}, g) : g \in G\} + \varepsilon \quad (6.1.2)$$

Since G is compact, there exists a subsequence $< g_{n_i} >$ of $<g_n>$ such that $<g_{n_i}> \rightarrow g_0$ and so (6.1.2) implies

$$d(x_1^{(0)}, +g_0) + d(x_2^{(0)}, g_0) \leq \inf \{d(x_1^{(0)}, g) + d(x_2^{(0)}, g): g \in G\} + \varepsilon$$

i.e. $g_0 \in P_{G(\varepsilon)}(x_1^{(0)}, x_2^{(0)})$.

Since $<g_n> \in N$ and N is closed, $g_0 \in N$. Consequently,

$g_0 \in P_{G(\varepsilon)}(x_1^{(0)}, x_2^{(0)}) \cap N$ i.e. $(x_1^{(0)}, x_2^{(0)}) \in B$. Thus B is closed and so $P_{G(\varepsilon)}$ is upper semi-continuous.

The next result deals with the structure of the sets $P_{G(\varepsilon)}(x_1, x_2)$

Proposition 6.1.5 [76]. Let G be a non-empty compact subset of metric space (X, d) and $P_{G(\varepsilon)} : X \times X \to 2^G$ ($\equiv$ the collection of all bounded subsets of G) be the ε-simultaneous approximation map of XxX into G defined $P_{G(\varepsilon)}(x_1, x_2) = \{g_0 \in G : d(x_1, g_0) + d(x_2, g_0) \le r + \varepsilon\}$

where $r = \inf\{d(x_1, g) + d(x_2, g) : g \in G\}$.

Then the set $P_{G(\varepsilon)}(A \times A) = \cup\{P_{G(\varepsilon)}(x_1, x_2) : x_1, x_2 \in A\}$ is compact for any compact subset A of X.

Proof. Let $< g_{n_i} >$ be any sequence in $P_{G(\varepsilon)}(A \times A) \subseteq G$. Since G is compact, there exists a subsequence $< g_{n_i} >$ of $< g_n >$ such that $< g_{n_i} > \to g_o \in G$

Since $< g_{n_i} >$ is a sequence in $P_{G(\varepsilon)}(A \times A)$, there exist

$<x_1^{(n_i)}, x_2^{(n_i)}>$ in AxA such that for each n_i,

$$g_{n_i} \in P_{G(\varepsilon)}(x_1^{(n_i)}, x_2^{(n_i)})$$

i.e. $d(x_1^{(n_i)}, g_{n_i}) + d(x_2^{(n_i)}, g_{n_i}) \le \inf\{d(x_1^{(n_i)}, g) + d(x_2^{(n_i)}, g): g \in G\} + \varepsilon$ (*)

Since AxA is compact $<x_1^{(n_i)}, x_2^{(n_i)}>$ has a subsequence

$<x_1^{(n_{ij})}, x_2^{(n_{ij})}> \to (x_1^{(0)}, x_2^{(0)}) \in AxA$. Since

$\inf\{d(x_1^{(n_{ij})}, g) + d(x_2^{(n_{ij})}, g) : g \in G\} \to \inf\{d(x_1^{(0)}, g) + d(x_2^{(0)}, g): g \in G\}$

and

$d(x_1^{(0)}, g_{n_{ij}}) + d(x_2^{(0)}, g_{n_{ij}}) \le d(x_1^{(0)}, x_1^{(n_{ij})}) + d(x_1^{(n_{ij})}, g_{n_{ij}}) +$

$$d(x_2^{(0)}, x_2^{(n_{ij})}) + d(x_2^{(n_{ij})}, g_{n_{ij}})$$

$$\le d(x_1^{(0)}, x_1^{(n_{ij})}) + d(x_1^{(n_{ij})}, g_{n_{ij}}) +$$

$$\inf\{d(x_1^{(n_{ij})}, g) + d(x_2^{(n_{ij})}, g) : g \in G\} + \varepsilon \quad \text{(by (*))}.$$

we have

$$\lim [d ((x_1^{(n_{i_j})}, g_{n_{i_j}}) + d (x_1^{(n_{i_j})}, g_{n_{i_j}}))] \leq \inf (d \{x_1^{(0)}, g) + d (x_2^{(0)}, g) : g \in G\} + \varepsilon$$

i.e. $d (x_1^{(0)}, g_0) + d (x_2^{(0)}, g_0) \leq \inf (d \{x_1^{(0)}, g) + d (x_2^{(0)}, g) : g \in G\} + \varepsilon$

i.e. $g_0 \in P_{G(\varepsilon)}(x_1^{(0)}, x_2^{(0)}) \subset P_{G(\varepsilon)}(A \times A)$. Hence $P_{G(\varepsilon)}$ (AxA) is compact.

6.2   Best Simultaneous Co-Approximation

This section deals with the problem of best simultaneous co-approximation in metric linear spaces and convex metric spaces.

To start with, we recall a few definitions.

Definition 6.2.1. Let (X, d) be a metric space and G a non-empty subset of X. An element $g_0 \in g$ is called a **best simultaneous co-approximation** to x, y $\in$ X from G if

$$d (g_0, g) \leq \max \{d (x, g)\} d (y, g) \quad \text{for all } g \in G.$$

The set of all best simultaneous co-approximations to x, y $\in$ X from G is denoted by $S_G(x, y)$. G is called an **existence set** if $S_G(x,y)$ contains at least one element, G is called a **uniqueness set** if $S_G(x,y)$ contains atomst one element and G is called an **existence and uniqueness** set if $S_G(x, y)$ contains exactly on element.

Definition 6.2.2. A metric linear space (X,d) is said to be **totally complete** [50] if it has the property that its d-bounded closed sets are compact.

Every totally complete metric linear space is finite dimensional but a finite-dimensional metric linear space need not be totally complete [50]. However, finite dimensional normed linear spaces are totally complete.

Some properties of the set $S_G(x, y)$ are as under:

Lemma 6.2.1. If G is a subset of a metric space (X,d) and x, y $\in$ X, then

(a)   if $g_0 \in S_G(x,y)$ then for every $g \in G$

   $d (x, g_0) \leq 2 \max \{ d (x, g), d (y, g) \}$

   $d (y, g_0) \leq 2 \max \{ d (x, g), d (y, g) \}$,

(b) $S_G(x,y)$ is bounded,

(c) $S_G(x,y)$ is closed if G is closed.

Proof.

(a)   is easy to verity

(b)   Let $g_0 \in S_G(x,y)$. Then by part (a), we have

$$d(x, g_0) \leq 2 \max \{ d(x, g), d(y, g) \}$$

for all $g \in G$ and so

$$d(g_0, x) \leq 2 \inf_{g \in G} \max \{ d(x, g), d(y, g) \}$$

$$\equiv 2 d(x, y; G).$$

Then for arbitrary g_0, $g_0' \in S_G(x,y)$

$$d(g_0, g_0') \leq d(g_0, x) + d(x, g_0')$$

$$\leq 4d(x, y; G)$$

implying thereby that $S_G(x,y)$ is bounded.

(c)   Let $\{ g_n \}$ be any sequence of element of $S_G(x,y)$ such that $\{ g_n \} \dashrightarrow g_0$.

Since G is closed, $g_0 \in G$. For any $g \in G$,

Consider

$$d(g_0, g) \quad \leq d(g_0, g_n) + d(g_n, g)$$

$$\leq (g_0, g_n) + \max \{ d(x, g), d(y, g) \}$$

$$\rightarrow o + \max \{ d(x, g), d(y, g) \} \text{ as } n \rightarrow \infty .$$

Thus $g_0 \in S_G(x,y)$ and so $S_G(x,y)$ is closed.

For normed linear spaces Lemma 6.2.1 was proved in [53] – proposition 3.1.

Note : The following results can be easily verified :

(i)   For $g_0 \in S_G(x,y)$), $\max d(g_0, x), d(g_0, y)\} \leq 2\max\{d(x,g), d(y,g)\}$

(ii)   For $g_0 \in P_G(x, y), d(g_0, g) \leq 2 \max \{d(x,g), d(y,g)\}$

(iii)   For $g_0 \in P_G(x), d(g_0, g) \leq 2 d(x,g)$

(iv)   For $g_0 \in R_G(x), d(g_0, x) \leq 2d(x,g)$

for all $g \in G$.

(V) For x∈ X\G and $g_0 \in G$ if g_0 is a best co-approximation to x from G, then g_0 is a best simultaneous co-approximation to x, y from G, for every y∈ X\G. But, if g_0 need not be a best co-approximation to either x or y. Therefore, $R_G(x) \cup R_G(y) \subset S_G(x,y)$.

It was proved in [8] that if G is a convex subset of a convex metric space (X,d), x, y∈ X and g_1, g_2 are best simultaneous approximations to x and y by the elements of G then $W(g_1, g_2, \lambda) \in G$ is also a best simultaneous approximation to x, y. The following theorem shows that the same is true in case of best simultaneous co-approximation:

Theorem 6.2.1. [82]. If G is a convex subset of a convex metric space (X, d) and x, y∈ X. Then $S_G(x,y)$ is convex.

Proof. Let $g_1, g_2 \in S_G$(x,y) and $\lambda \in [0,1]$.

Then $W(g_1, g_2, \lambda) \in G$ as $g_1, g_2 \in G$ and G is a convex set.

Consider

$$d(W(g_1, g_2, \lambda), g) \leq \lambda d(g_1, g) + (1-\lambda) d(g_2, g)$$
$$\leq \max\{d(g_1, g), d(g_2, g)\} \qquad (6.2.1)$$

Since $g_1, g_2 \in S_G$(x,y),

$$d(g_1, g) \leq \max\{d(x,g), d(y, g)\}, \text{ and} \qquad (6.2.2)$$
$$d(g_2, g) \leq \max\{d(x,g), d(y, g)\}.$$

Now (4.2.1) and (4.2.2) imply

$$d(W(g_1, g_2, \lambda), g) \leq \max\{d(x,g), d(y, g)\}$$

for every $g \in G$ and so $W(g_1, g_2, \lambda) \in S_G$(x,y).

It was proved by Diaz and McLaughlin [20] that for a normed linear space X, a finite dimensional subspace G of X and x, $y \in X/G$, if $g_0 \in G$ is a best approximation to (x+y)/2 from G, then g_0 is not a best simultaneous approximation to x, y from G. However, in case of best co-approximation we have:

Theorem 6.2.2 [82]. Let (X, d) be a convex metric space, G a subset of X, $x, y \in X \setminus G$, if $g_0 \in G$ and $0 \leq \alpha \leq 1$. If g_0 is a best co-approximation to W

(x, y, α) for some α, then g_0 is a best simultaneous co-approximation to x, y from G.

Proof. Assume that g_0 is a best co-approximation to W (x, y, α) for some $\alpha \in [0,1]$. Then for every $g \in G$ it follows that

$$d(g_2, g) \leq d(W(x,y,\alpha),g)$$
$$\leq \alpha\, d(x,g) + (1-\alpha)\, d(y, g)$$
$$\leq \max\{d(x, g), d(y, g)\}.$$

Thus g_0 is a best simultaneous co-approximation to x, y for G.

Remark. For normed linear spaces this result was proved in [64] – Theorem 4.4

For a convex metric linear space (X, d) and x, y belonging to convex set G in X, $\alpha x + (1-\alpha)y$, $0 \leq \alpha \leq 1$ is a best simultaneous co-approximation to x, y from G. As for any $g \in G$

$$d(\alpha x + (1-\alpha)y, g) \leq \alpha d(x, g) + (1-\alpha)d(y, g) \quad \text{(by the convexity of (X,d))}$$
$$\leq \max\{d(x, g), d(y, g)\}$$

Also every element belonging to $S_G(x, y)$ is of the form $\alpha x + (1-\alpha)y$, $0 \leq \alpha \leq 1$.

Hence for a convex metric linear space (X, d) and a convex subset G of X, if x, $y \in G$ then

$$S_G(x,y) = \{\alpha x + (1-\alpha) y : 0 \leq \alpha \leq 1\}.$$

In the next theorem, we list some more properties of the set $S_G(x,y)$ in metric linear spaces.

Theorem 6.2.3 [82]. Let (X, d) be a metric linear space, G a subspace of X and x, $y \in X$. Then the following results hold:

(i) $\quad S_G(x + g, y+g) = S_G(x, y) + g$ for $g \in G$.

(ii) $\quad S_G(\alpha x, \alpha y) = \alpha S_G(x, y)$ for x, $y \in G$, $\alpha \in R$

The prof of Theorem 6.2.3 is a minor modification of that of Proposition 3.3 given in [53], for normed linear spaces.

The next result gives another property of the set $S_G(x, y)$ in totally complete metric linear spaces.

Theorem 6.2.4 [82]. Let (X, d) be a totally complete metric linear space, G a non-empty closed subset of X. Then $S_G(x, y)$ is compact.

Proof. Since $S_c(x, y)$ is closed and bounded (lemma 6.2.1) and the space is totally complete. $S_G(x, y)$ is compact.

The following theorem proves the upper semi-continuity of the mapping S_G for totally complete metric linear spaces:

Theorem 6.2.5 [82] Let (X, d) be a totally complete metric linear space (X, d). Then the set-valued map $S_G\{x, y) : x, y \in X\} \rightarrow 2^G$ is upper semi-continuous.

Proof. Let A be a closed subset of G, we have to show that $B = \{(x,y) : x, y \in X, S_G(x,y) \cap A \neq \phi\}$ is a closed subset of X. Let $<(x_n, y_n)>$ be a sequence in B such that $<(x_n, y_n)> \rightarrow (x_0, y_0)$ for some $(x_0, y_0) \in X$. Since $(x_n, y_n) \cap A$ is non-empty, choose

$g_n \in S_G(x_n, y_n) \cap A$ for each n.

Consider the set C=closure of the set $\{g_1, g_2, ..., g_n, ...\}$. Using Lemma 6.2.1 (a), it is easy to see that the set C is a bounded set. Since C is a closed and bounded subset of the totaly complete space G, C is compact. Therefore the sequecne $<g_n>$ has a subsequence $\{g_{n_k}\}$ converging to g_0. Since A is closed, $g_0 \in A$.

Now to prove that B is closed, it is sufficient to prove that $g_0 \in S_G(x_0, y_0)$. For every $g \in G$.

$$d(g, g_0) \quad \leq \quad d(g, g_{n_k}) + d(g_{n_k}, g_0)$$
$$\leq \quad \max\{d(x_{n_k}, g), d(y_{n_{pk}}, g)\} + d(g_{n_k}, g_0)$$
$$\rightarrow \quad \max\{d(x_0, g), d(y_0, g) + 0 \text{ as } n \rightarrow \infty.$$

Therefore $g_0 \in S_G(x_0, y_0) \cap A$ and so $(x_0, y_0) \in B$ i.e. B is closed

--- X ---

84

CHAPTER-VII
FIXED POINTS AND APPROXIMATION

This chapter deals with the structure of fixed point set, the problem of invariant approximation and an application of fixed point theorem to ε-simultaneous approximation.

It is known (see e.g. [12] Theorem 6, p. 243) that for a closed convex subset K of a strictly convex normed linear space X and a non-expansive mapping $T:K \to K$ the fixed point set (possibly empty) of T is a closed convex set. We extend this result to pseudo strictly convex metric linear spaces in the first section.

Fixed points of non-expansive mappings have been extensively discussed in strictly convex normed linear spaces (see e.g. [36]. Using fixed point theory, Meinardus [43] and Brosowski [14] established some interesting results on invariant approximation in normed linear spaces. Later various researchers obtained generalizations of their results (see e.g. [36] and the references cited therein). In the second section we extend and generalize the work of Brosowski [14], Hicks and Humphries [31], Khan and Khan [39] and Singh [86], [87] to metric spaces having convex strucrure and to metric linear spaces having strictly monotone metric (a notion introduced by Guseman and Peters [29]). We have proved the existence of an invariant point x_0 in the set $P_C(x)$ satisfying certain conditions. We have also established a result on invariant approximation in strictly convex metric spaces in this section.

Some applications of fixed point theorems to best simultaneous approximation were given by Ismat Beg and Naseer Shahzad [9], R.N. Mukherjee and V. Verma [45] and few others when the underlying spaces are normed linear spaces. Using a result of Beg and Azam [7] on fixed points of mutivalued mappings, we give an application of a fixed point theorem to ε-simultaneous approximation. when the spaces are convex metric spaces in the third section of this chapter.

7.1   Fixed Points in Pseudo Strictly Convex Spaces

The following theorem give the structure of the fixed point set of a non-expansive mapping in pseudo strictly convex metric spaces:

Theorem 7.1.1 [78] : Let K be a closed convex subset of a convex metric linear space (X, d) with pseudo strict convexity and $T : K \rightarrow X$, a non-expansive mapping. Then fixed point set (possibly empty) of T is a closed, convex set.

Proof: Let $F = \{x \in K: Tx=x\}$ be the fixed point set of T. Firstly we prove the closedness of the set F. Let x be a limit point of F. then there exists a sequence $\{x_n\}$ in F such that $\{x_n\} \rightarrow x$. Since a non-expansive mappint is always continuous, we get $Tx_n \rightarrow Tx$. Also $Tx_n = x_n \rightarrow x$. as $x_n \in F$ and so Tx=x i.e. $x \in F$. Hence F is a closed set.

Now we show that F is convex. Let x, $y \in F$ and $\lambda \in [0, 1]$ Then x, $y \in K$ $y \in K$ and so $\lambda x + (1-\lambda)$ y=z (say) $\in K$ Consider

$$d(x, Tz) = d(Tx, Tz)$$
$$\leq d(x, z) \text{ (as T is non-expansive)}$$
$$= d(x, \lambda x+(1-\lambda) y)$$
$$\leq \lambda d(x, x) + (1-\lambda) d(d, y)$$
$$\text{(by the convexity of X)}$$
$$= (1-\lambda) d(x, y).$$

Also, $d(Tz, y) = d(Tz, Ty)$
$$\leq d(z, y) \text{ (as T is non-expansive)}$$
$$= d(\lambda x + (1-\lambda) y, y)$$
$$\leq \lambda d(x, y) + (1-\lambda) y, y)$$
$$\text{(by the convexity of X)}$$
$$= \lambda d(x, y).$$

Therefore, $d(x, Tz) + d(Tz, y) \leq d(x, y)$.
Also by the triangle inequality,
$$d(x, y) \leq d(x, Tz) + d(Tz, y).$$
Therefore
$$d(x, y) = d(x, Tz) + d(Tz, y)$$
i.e. $d((x-Tz) + (Tz-y), 0)=d(x-Tz, 0) +d(Tz-y, 0)$.
So by the pseudo strict convexity of X, we have
x-Tz=k (Tz-y) i.e. Tz=x/(1+k)+ky/(1+k) i.e. $Tz \in [x, y]$

86

Next we show that the non-expansivity of T implies Tz=z. Since z∈ [x, y] and Tz∈ [z, y], z will be either between x and Tz or between x and Tz or between Tz and y.

Suppose z lies between x and Tz then

d (Tz, x)   =   d (Tz, Tx) as x∈ F

             ≤   d (z, x) as T is non-expansive.

Now d (Tz, z)   +   d (z, x) = d (Tz, x) (as z∈ [x, Tz])

             ≤   d (z, x)

implies d (Tz, z) = 0 and so Tz = z

Similarly, if z is between Tz and y, we shall get Tz=z. Therefore, Tz = z i.e. z ∈ F i.e. λ x + (1-λ) y∈ F, for all x, y∈ F and $0 \leq \lambda \leq 1$. Hence F is a convex set.

7.2   Invariant Approximation

In this section, we extend and generalize some of the results of Brosowski [14], Hicks and Humphries [31], Khan and Khan [39] and Singh [86], [87] on invariant approximation in strictly convex metric spaces, in metric linear spaces having strictly monotone metric and in ε – chainable convex metric spaces.

To start with we recall a few definitions.

Defintion 7.2.1 [29]. Let (X, d) be a metric linear space. The metric d for X is said to be **strictly monotone** [29] if $x \neq 0, \ 0 \leq t < 1$ imply d (tx, 0) <d (x, 0).

Definition 7.2.2. A metric linear space (X, d) is said to satisfy **property (*)** if

$$d (\lambda x + (1-\lambda) y, z) \leq \lambda d (x, z) + (1-\lambda) d (y, z)$$

for every x, y, z ∈ X and $o \leq \lambda \leq 1$.

Clearly, every normed linear space satisfies property (*).

The following lemma in metric linear spaces satisfying property (*) will be used in the proof of Theorem 7.2.1.

Lemma 7.2.1 [79]Let (X, d) be a metric linear space satisfying property (*) , C a subset of X and x∈ X. Then $P_C(x) \subset \partial C \cap C$ where ∂C is the boundary of C.

Proof. Let $y \in P_C(x)$. For each positive integer n, let $\lambda_n = n/(n+1)$. Since $d(y, \lambda_n y + (1-\lambda_n)x) \leq (1-\lambda_n) \, d(x, y)$ for all n (using property (*)), $\lim_{n \to \infty}[\lambda_n y + (1-\lambda_n) x] = y$. So each neighbourhood of y contains atleast one $\lambda_n y + (1-\lambda_n) x$. Also, $d(y, \lambda_n y + (1-\lambda_n)x) \leq \lambda_n d(y,x) < d(y, x)$ for all n implies that $\lambda_n y + (1-\lambda_n) x \notin C$ for any n i.e. y is not an interior point of C and so $y \in \partial C$. Also $y \in P_C(x)$ implies $y \in C$.. Thus, $y \in \partial C \cap C$ and hence $P_C(x) \subset \partial C \cap C$.

We have the following result on invariant approximation in metric linear spaces:

Theorem 7.2.1 [79]. Let (X, d) be a metric linear space with strictly monotone metric d and C a subset of X. Let T be a non-expansive mapping on $P_C(x) \cup \{x\}$ where x is a T-invariant point. Then there is a x_0 in $P_C(x)$ which is also T-invariant provided.

(a) $T : \partial C \dashrightarrow C$

(b) $P_C(x)$ is nonempty, starshaped and compact.

(c) Either C is closed or (X, d) satisfies property (*)

Proof. Let P be starcentre of $P_C(x)$, then $\lambda x + (1-\lambda)p \in P_C(x)$ for every $x \in P_C(x).0 \leq \lambda \leq 1$. We claim that $T: P_C(x) \to P_C(x)$.

Suppose (X, d) satisfies property (*) then by Lemma 5.2.1, $P_C(x) \subset \partial C \cap C$. So for $y \in p_C(x)$, we get $Ty \in C$ as $T\partial C \dashrightarrow C$.

Suppose C is closed then $y \in P_C(x)$ implies $y \in \partial C$ leading to $Ty \in C$ as $T : \partial C \dashrightarrow C$.

Thus in both the cases, $Ty \in C$. Consider

$$d(x, Ty) = (Tx, Ty) \text{ (as x is a T-invariant point)}$$
$$\leq d(x, y) \text{ (as T is non-expansive on } P_C(x) \cup \{x\})$$
$$= d(x, C)$$
$$\leq d(x, Ty).$$

88

This give d (x, Ty) = d (x, C) i.e. Ty $\in P_C(x)$ for $y \in P_C(x)$ and hence T: $P_C(x)$ $\rightarrow P_C(x)$.

Let k_n, $0 \leq k_n < 1$ be a sequence of real numbers such that $k_n \text{---} > 1$ as $n \text{---} > \infty$. Define T_n: $P_C(x) \rightarrow P_C(x)$ as $T_n y = k_n Ty + (1-k_n) p$ for every $y \in p_C(x)$. Since T maps $p_C(x)$ into $P_C(x)$ for each n ($y \in p_C(x)$ and T: $P_C(x) \rightarrow P_C(x)$ imply Ty $\in P_C(x)$ and $P_C(x)$ being starshaped w.r.t. p we get $k_n Ty + (1-k_n) p \in P_C(x)$).

Also,

$$
\begin{aligned}
d (T_n x, T_n y) &= d (k_n Tx + (1-k_n)p, k_n Ty + (1-k_n)p) \\
&= d (k_n Tx, k_n Ty) \\
&= d (k_n (Tx\text{-}Ty), 0) \\
&< d (Tx\text{-}Ty, 0) \text{ (as d is strictly monotone)} \\
&= d (Tx, Ty) \\
&\leq d (x, y) \text{ (as T is non-expansive on } P_c(x) \cup \{x\})
\end{aligned}
$$

Hence T_n is non-expansive on $P_C(x) \cup \{x\}$ for each n. Since $P_C(x)$ is compact and starshaped, T_n has a unique fixed point, say, x_n for each n ([27], Theorem 2) i.e. $T_n x_n = x_n$ for each n.

Since $P_C(x)$ is compact, $<x_n>$ has a convergent subsequence $x_{n_i} \text{ --- } > x_0 \in P_C(x)$. We claim that $Tx_0 = x_0$.

Consider $x_{n_i} = T_{n_i} x_{n_i} = k_{n_i} + (1-k_{n_i}) p$. Taking limit as i$\text{---}>\infty$, we get $x_0 = x_0$ ($x_{n_i} \text{ --- } > x_0$ implies $T_{n_i} \text{ --- } > Tx_0$ as T is continuous on $P_C(x) \cup \{x\}$ i.e. $x_0 \in P_C(x)$ is T-invariant.

Since every normed liner space is a metric linear space with property (*) and the metirc induced by the norm is strictly monotone, we have:

Corollary 7.2.1 [14] . Let T be a non-expansive linear operator on a normed linear space X. Let C be a T-invariant subset of X and x a T-invariant point. If the set of best C-approximants to x is non-empty, compact and convex, then it contains a T-invariant point.

Corollary 7.2.2 [86] . Let T be a non-expansive mapping on a normed linear space X. Let C be a T-invariant subset of X and x_0 a T-invariant point in X. If D, the set of best C-approximants to x_0 is non-empty, compact and starshaped, then it contains a T-invariant point.

Corollary 7.2.3 [87] . Let X be a normed linear space and T:X--->X a mapping. Let C be a subset of X such that C is T-invariant and let x_0 be a T-invariant point in X. If D, the set of best C-approximants to x_0 is non-empty, compact and starshaped and T is

(i)   continuous on D

(ii) $\|x\text{-}y\| \leq d(x_0, C) \Rightarrow \|Tx\text{-}Ty\|$ for x, y in $D \cup \{x\}$,

then it contains a T-invariant point which is a best approximation to x_0 in C.

Note: The continuity of T on D need not be assumed as it follows from (ii)

Since each p-norm generates a translation invariant metric d satisfying property (*) and is strictly monotine, we have:

Corollary 7.2.4 [39]. Let $(E, \|\cdot\|_p)$ be a p-normed space, T:D-->E a non-expansive mapping with a fixed point $u \in E$ and C a closed T-invariant subset of E such that T is compact on C. If $P_C(u)$ is starshaped, then there exists an element in $P_C(u)$ which is also a fixed point of T.

In strictly convex metric spaces, we have the following result on invariant approximation:

Theorem 7.2.2 [79]. Let (X, d) be a strictly convex metric space and T a non-expansive mapping on $P_C(x) \cup \{x\}$ where x is a T-invariant point. Let C be a subset of X, $T: \partial C$--->C and $P_C(x)$ be non-empty and starshaped with starcentre q. Then $P_C(x) = \{q\}$ with Tq=q.

Proof. Let $p \neq q \in P_C(x)$. Then d (x, p) = d (x, q) = d (x, C). Since $p \neq q$, strict convexity of the space implies d (x, W (p, q, λ))<

90

dist (x,C) and so $W(p, q, \lambda) \notin P_C(x)$, $0 \le \lambda \le 1$. Starshapedness of $P_C(x)$ therefore implies p=q i.e. $P_C(x)=\{q\}$. Since X is convex, $P_C(x) \subset \partial C \cap C$ (Lemma 3.2 [91]). So for $y \in P_C(x)$, we get $Ty \in C$ as T:∂C --C. Consider

$$
\begin{aligned}
d\,(x,\, Ty) \quad &= \quad d\,(Tx,\, Ty) \quad \text{(as x is a T-invariant point)} \\
&\le \quad d\,(x,\, y) \quad \text{(as T is non-expansive on } P_C(x) \cup \{x\}\text{)} \\
&= \quad d\,(x,\, C) \\
&\le \quad d\,(x,\, Ty).
\end{aligned}
$$

This gives $d\,(x,\, Ty) = d\,(x,\, C)$ i.e. $Ty \in P_C(x)$ for $y \in P_C(x)$ and so $T: P_C(x) ----> P_C(x)$. Hence $Tq \in P_C(x) = \{q\}$, i.e. Tq=q.

7.3   Fixed Points and ε- Simultaneous Approximation

Using a result of Beg and Azam [6] on fixed point of multivalued mappings, we give an application of a fixed point theorem to ε- Simultaneous approximation in convex metric spaces in this section. We start with a few definitions.

Defintion 7.3.1. A metric space (X, d) is said to be ε-**chainable** (see e.g. [9]) if given x , y$\in$ X there is an ε-chain from x to y (i.e. a finite set of points x=z_0, z_1, z_2,...,z_n=y such that $d(z_{i-1}, z_i) < \varepsilon$ for all i= 1,2, ..., n.

Definition 7.3.2. A mapping T:X --> CB(X) is called an (ε, λ)- uniformly locally contractive mapping (where ε>0 and 0<λ <1) (see e.g.[9] if x, y$\in$X and d(x, y) < ε then H (Tx, Ty) $\le \lambda d$ (x, y) where H stands for Hausdorff metric on CB(x).

An application of a fixed point theorem to b.s.a. was given by Beg and Shahzad [9]. Using the following simplified version of a result due to Beg and Azam [7]:

Lemma 7.3.1 Let (x,d) be a complete ε-chainable metric space T:X -->CB(X) satisfies the condition:

$0 < d(x, y) < \varepsilon$ implies H (Tx, Ty) <kd (x, y)

where $k \in [0,1[$, then there exists a fixed point of T.

We now extend the result of [9] to ε-s.a.

Theorem 7.3.1 [77]. Let (x, d) be an ε-chainable convex metric space satisfying condition (I) and $T : X \to CB(X)$ be a multivalued mapping. Let $G \in CB(X)$. For $F \in CB(X)$, if $cent_G(F, \varepsilon)$ is compact, starshaped, T-invariant and T is

(i)   continuous on $cent_G(F, \varepsilon)$ and

(ii) $d(x, y) \leq H(F, G)$ implies $H(Tx, Ty) \leq d(x, y)$ for all   x, y in $cent_G(F, \varepsilon) \cup F$, then $cent_G(F, \varepsilon)$ contains a T-invariant point.

Proof. The proof of this theorem is a minor modification of the one given by beg and Shahzad [9] for b.s.a. in normed linear spaces.

Let p be the starcentre of $cent_G(F, \varepsilon)$ then $W(x, p, \lambda) \in cent_G(F, \varepsilon)$ for each $x \in cent_G(F, \varepsilon)$. Let $<k_n>$ be a sequence of real numbers with $0 \leq k_n < 1$ converging to 1. Define $T_n : cent_G(F, \varepsilon) \to CB(cent_G(F, \varepsilon))$ as

$$T_n x = W(Tx, p, k_n) = \bigcup_{y \in Tx} W(y, p, k_n)$$

for all x in $cent_G(F, \varepsilon)$.

Clearly T_n is well defined as for $g_0 \in cent_G(F, \varepsilon)$ we have

$d_F(W(Tg_0 , p, kn)) = d_F(W(g_0 , p, kn))$ as $g_0 \in cent_G(F, \varepsilon)$ is T-invariant

$\leq D(F, G) + \varepsilon$ as $cent_G(F, \varepsilon)$ being starshaped,

$\qquad\qquad W(g_0, p, kn) \in cent_G(F, \varepsilon)$ i.e. for $g_0 \in cent_G(F, \varepsilon)$, $T_n g_0 = W(Tg_0, p, k_n) \in CB(cent_G(F, \varepsilon))$.

Using condition (I), we have

$H(T_n x, T_n y) = H(W(Tx, p, k_n), W(Ty, p, k_n))$

$\qquad\qquad \leq k_n H(Tx, Ty)$

$\qquad\qquad \leq k_n d(x, y)$

for all $d(x, y) \leq H(F, G)$ which implies that T_n is a $(H(F, G), k_n)$ uniformly local contraction for each n=1, 2, 3, ... It follows by Lemma 5.3.1 that each T_n has a fixed point, say, x_n. Since $cent_G(F, \varepsilon)$ is given to be compact, $\{x_n\}$ has a convergent subsequence $x_{n_i} \to z$ and so continutiy of T on $cent_G(F, \varepsilon)$ implies $Tx_{n_i} \to Tz$.

$$x_{n_i} \in Tn_i x_{n_i} = W(Tx_{n_i}, p, k_{n_i})$$

Since $k_{n_i} \to 1$, $z \in Tz$ (as $k_{n_i} \to 1$ implies $z \in W(Tz, p, 1) = Tz$).

--- X ---

CHAPTER –VIII

ON NON-EXPANSIVE RETRACTS

In order to generalize a theorem of Belluce and kirk [10] on the existence of a common fixed point of a finite family of commuting non-expansive mappings, Ronald E. Bruck Jr.. [17] studied some properties of fixed-point sets of non-expansive mappings in Banach spaces. In this paper, we extend some results of [17] to convex metric spaces. We also prove that the fixed point set of a non-expansive mapping satisfying conditional fixed point property (CFP) is a non-expansive retract and hence metrically convex.

To start with, we recall a few definitions.

Definition 8.1 [17]. Let (X,d) be a metric space and C a closed convex subset of X. A mapping $T:C \to X$ is said to satisfy conditional fixed point property (CFP) if either T has no fixed point or T has a fixed point in every non-empty bounded closed convex [17].

Definition 8.2 [17] A set C is said to have the **fixed point property (f.p.p.)** for non-expansive mappings if every non-expansive mapping of C into C has a fixed point

Definition 8.3. A subset of a metric space (x,d) is said to be **matrically convex** [41] if for each pair of distinct points x_0 , x_1 of F, there exists a point y in F distinct from x_0 and x_1 such that $d(x_0 \ x_1) = d(x_0 \ y) + d(y, x_1)$.

The following theorem shows that the non-expansive retracts are metrically convex.

Theorem 8.1 [81]. Let C be a non-empty convex subset of a convex matric space (X,d). Then non-expansive retracts of C are metrically convex.

Proof. Let r be a non-expansive retraction of C onto F and x_0 , x_1 be two distinct points of F. For t in [0,1], consider $W(x_0 \ x_1 ,t) = x_t \in C$ Since $r(x_0) = x_0 \neq x_1$, by the continuity of W and r the mapping $t \to r(x_t)$ is distinct from x_0 and x_1 .

Let y be one such r (x_t). Then $y \in F$ is distinct from x_0 and x_1 while

$d(x_0, x_1)$ $= d(r(x_0), r(x_1))$

$\leq d(r(x_0), r(x_t)) + d(r(x_t), r(x_1))$

$\leq d(x_0, x_t) + d(x_t, x_1)$ (r being non-expansive)

$= d(x_0, w(x_0, x_1, t)) + d(w(x_0, x_1, t), x_1)$

$\leq t\, d(x_0, x_0) + (1-t)\, d(x_0, x_1) + t\, d(x_0, x_1)$

$+(1-t)d(x_1, x_1)$ (by the convexity of X)

$= d(x_0, x_1).$

So, equality holds throughout and in particular,

$d(r(x_0), r(x_1)) = d(r(x_0), r(x_t)) + d(r(x_t), r(x_1))$

i.e. $d(x_0, x_1) = d(x_0, y) + d(y, x_1)$. Hence F is metrically convex.

Next we shall show that the fixed point set of a non-expansive mapping satisfying (CFP) is a non-expansive retract of C and hence metrically convex. To achieve this, we firstly prove a few lemmas:

Lemma 8.1. [81]. Suppose F is a non-empty subset of a locally compact set C in a metric space (X,d) and let $N(F) = \{f: f:C \to C$ is non-expansive and $fx = x$ for all $x \in F$ $\}$. Then $N(F)$ is compact i.e. the set of all non-expansive retractions of C onto F is compact.

Proof. Fix $x_0 \in F$ and for $x \in C$ define

$C_x = \{y \in C : d(y, x_0) \leq d(x, x_0))$.

For each x in C and f in $N(F)$, $f(x) \in C_x$ since $f(C) \subset C$ and

$d(fx , x_0) = d(fx , fx_0) \le d (x , x_0)$.

Thus by the definition of cartesian product of indexed family, $N(F)$ is a subset of the cartesian product $P = \coprod_{x \in C} C_x$.

Since C is locally compact, each C_x is compact and so by Tychonoff's theorem P is compact.

Now we show that $N(F)$ is closed in P. Let $<f_n>$ be a sequence of elements in $N(F)$ with $<f_n> \to f_0$ i.e. $f_n : C \to C$ is non-expansive and $f_n x = x$ for all $x \in F$ and for each n.

Then. $f_n x = x \Rightarrow \lim_{n \to \infty} f_n x = x \Rightarrow f_0 x = x.$

Also,

$$d(f_0 x, f_0 y) = d(\lim_{n \to \infty} f_n x, \lim_{n \to \infty} f_n y)$$

$$= \lim_{n \to \infty} d(f_n x, f_n y)$$

$$\le d(x,y) \quad (\text{each} f_n \quad \text{being non-expansive})$$

Therefore, $N(F)$ is closed in P and hence compact.

Lemma 8.2. [81] , Suppose F is a non-empty subset of a locally compact set C in a metric space (X,d). Then there exists an r in $N(F) = \{f : f: C \text{---}>C$ is non-expansive and fx=x for all $x \in F\}$ such that each f in $N(F)$ acts as an isometery on the range of r i.e.

d(f(r(x)), f(r(y))) = d (r (x), r (y)) for all x , $y \in C$.

Proof. Define an order $\le$ on N (F) by setting f ≤ g if

d(fx,fy) $\le$ d(gx,gy) for all x,y in C with inequality holding for atleast one pair of x, y.

Clearly $\le$ is a partial order on N(F).

For each f in N(F), we define M={ $g \in N(F) : g \leq f$ }.

Clearly M is closed in N(F) (Let $<g_n>$ be a sequence of elements in M with $<g_n> \to g_0$. So $g_n : C \to C$ is non-expansive, $g_n x = x$ for all $x \in F$ and $g_n \leq f$. As discussed in Lemma 8.1 ; we have $g_0 : C \to C$ is non- expansive with $g_0 x = x$ and also $g_n \leq f$ implies $\lim_{n \to \infty} g_n \leq f$ and hence $g_n \in C$). Therefore M is closed in N(F) and hence compact as N(F) is compact.

It follows that (N(F), $\leq$) contains a minimal point. Indeed by zorn's lemma, it is sufficient to show that whenever $\{g_\lambda : \lambda \in \Lambda\}$ is linearly ordered by $\leq$, then there exists g in N(F) with $g \leq g_\lambda$ for all λ . But if the g_λ s are linearly ordered by $\leq$, family $\{M(g_\lambda) : \lambda \in \Lambda\}$ is linearly ordered by inclusion. Since each $M(g_\lambda)$ is compact and non-empty, there exists $g \in \cap M(g_\lambda)$ i.e. there exists g in N(F) with $g \leq g_\lambda$ for each λ .

Hence there exists a minimal element r of N(F) It si easily verified that for $f \circ r \in N(F)$ whenever $f \in N(F)$

(i) $d(f \circ r(x), f \circ r(y)) = d(f(r(x)), f(r(y)))$

$$\leq d(r(x), r(y))$$

and

(ii) $(f \circ r)(x) = f(r(x)) = f(x) = x$ or $f \circ r \in N(F)$.

Now in particular $d(f \circ r(x), f \circ r(y)) \leq d(r(x), r(y))$ \qquad (8.1)

If inequality holds in (6.1) for any pair x , y in C then $f \circ r < r$ while for $f \circ r \in$ N(F), contradicting the minimality of r in N(F). Therefore, equality holds in (8.1) for each x , y in C.

i.e. d(f (r (x)) , f (r (y))) = d (r (x), r (y)) for all x, y$\in C$.

Using Lemmas 8.1 and 8.2, non-expansive retracts of C can be described as:

Lemma 8.3 [81]. Suppose C is locally compact and F a non- empty subset of C in a metric space (X,d). Suppose that for each z in C there exists h in N(F) such that h(z) $\in$ F. Then F is a non-expansive retract of C.

Proof. Since F is a non-empty subset of a locally compact set C, by Lemma 8.2, there exists an r in N(F) = { f : f : C $\rightarrow$ C is non-expansive and fx=x for all x $\in$ F } such that each element of N(F) acts as an isometry on the range of r. Since r $\in$ N(F) we get r:C $\rightarrow$ C is non-expansive and rx=x for all x $\in$ F. So, to show that r is a non-expansive retraction of C onto F, it is sufficient to show that r(x) $\in$ F for each x in C.

Applying the given hypothesis to the point z=r(x), there exists h in N(F) with h (r(x)) $\in$ F. Let y =h(r(x)). By Lemma 8.2,

$$d(h(r(x)), h(r(y))) = d(r(x), r(y))$$ (8.2)

Since y $\in$ F and h, r $\in$ N (F) , we have h (y) = y and r (y) = y and hence y=h (r (y)) while y=h (r (x)) by definition. So it follows from (8.2) that r(x) = r (y) . Since r (y) = y, r(x) = y $\in$ F for each x $\in$ C. Therefore r : C $\rightarrow$ F is a non-expansive retraction and hence F is a non-expansive retract of C.

It was show in [16] that if C is a closed convex subset of the real, reflexive, strictly convex Banach space X and T: C $\rightarrow$ C is non-expansive then F(T) is a non-expansive retract of C. The same conclusion was drawn in [17] when C is a non-empty closed convex, locally weakly compact subset of a Banach space and T:C $\rightarrow$ C is a non-expansive mapping satisfying (CFP). In convex metric spaces we have.

Theorem 8.2 [81] Let C be a locally compact, convex set in a convex metric space (X,d) with properties (I) and (I^*) and T:C $\rightarrow$ C a non-expansive mapping satisfying (CFP) then F(T) is a non-expansive retract of C.

Proof. Since the empty set, by definition, is a non-expansive retract of C, we assume that F(T) $\neq$ ϕ.

Fix z in C and define k = { f (z) : f $\in$ N (F (T))).

98

Clearly K is the image of N(F(T)) under the zth co-ordinate projection map of P onto C_z. Since N(F(T)) is compact (Lemma 8.1) and the projection continuous, K must be compact and hence bounded. Clearly K is non-empty. We claim that K is convex.

Let f, g $\in$ N (F(T)) . and $0 \leq \lambda \leq 1$. Since N(F (T)) =

{ f : f : C $\rightarrow$ C is non-expansive and fx=x for all x $\in$ F (T)), We are to show that

(i)   W (f,g, λ) (x) = x for all x $\in$ F (T) and

(ii)   W (f, g, λ) : C $\rightarrow$ C is non-expansive

(i)   Let x $\in$ F (T) Consider

$$d(W(f, g, \lambda)(x), x) = d(W(f(x), g(x), \lambda), x)$$
$$\leq \lambda \ d (f(x), x) + (1-\lambda) d (g(x), x))$$
$$\text{(by the convexity of X)}$$
$$= \lambda \ d(x, x) = (1-\lambda) d(x,x) \text{ (as f, g} \in N (F(T)))$$
$$= 0$$

implying W(f , g , λ) (x) = x for all x $\in$ F(T). Consider

$$d(W(f,g, \lambda)(x), W(f,g, \lambda)(y))$$

$$= d(W(f(x), g(x), \lambda), W(f(y), g(y), \lambda))$$

$$\leq d(w(f(x), g(x), \lambda), w(f(x), g(y), \lambda))$$

$$+ d(w(f(y), g(y), \lambda), w(f(x), g(y), \lambda))$$

$$\leq (1-\lambda) d(g(x), g(y)) + \lambda d(f(y), f(x))$$

$$\text{(by properties } (I^*) \text{ and } (I))$$

$$\leq (1-\lambda) d(x, y) + \lambda d(x, y)$$

$$\text{(as f, g are non-expansive)}$$

$$= d(x, y)$$

implying that W (f , g , λ) is non-expansive. Therefore k is a bounded, non empty, closed and convex subset of C.

Since $T \circ f \in N (F (T))$ whenever $f \in N (F (T))$ as

$$d(T \circ f(x), T \circ f(y)) = d(T(f(x)), T(f(y)))$$

$$=d(Tx, Ty) \quad (\text{as } f (x) = x \text{ and } f (y) = y)$$

$$\leq d (x, y) \quad (\text{as T is non-expansive})$$

and $\qquad T \circ f(x) = T (f(x))$

$$= T (x) \quad (\text{as } f (x) = x)$$

$$= x \text{ for all } x \in F (T),$$

we have $T (K) \cap K$.

Since T satisfies (CFP) and K is a non-empty bounded closed convex set that T leaves invariant, T has a fixed point in K. So there exists h in N (F(T)) with h (z) in F(T).

Since this is so for each z in C, by Lemma 8.3, F(T) is a non expansive retract of C.

Corollary 8.1 [81]. Let C be a non-empty locally compact convex subset of a convex metric space (X,d) satisfying properties (I) and (I^*), T: C $\rightarrow$ C be non-expansive mapping satisfying (CFP). Then F (T) is metrically convex.

Proof. By Theorem 8.2, F (T) is a non expansive retract of C and hence by Theorem 8.1 , it is metrically convex.

The following lemma will be used in proving our next result.

Lemma 8.4 [81]. Suppose C is a convex subset of a strictly convex metric space (x,d) with properties (I) and (I^*) and T_1, T_2,: C $\rightarrow$ C be non-expansive mappings such that $F (T_1) \cap F (T_2) \neq \phi$. Then exists a non-expansive mapping T:C $\rightarrow$ C such that $F (T) = F (T_1) \cap F (T_2)$.

Proof. Define $T : C \rightarrow C$ as

$$T (x) = W(T_1, T_2, \lambda) (x) = W(T_1(X), T_2(X), \lambda) \text{ for } 0 \leq \lambda \leq 1.$$

We claim that T is the required mapping.

Consider

$$d(Tx, Ty) = d(W(T_1, T_2\lambda)(x), W(T_1, T_2, \lambda)(y))$$

$$d(W(T_1(x), T_2(x), \lambda), W(T_1(y), T_2(y), \lambda))$$

$$= d(W(T_1(x), T_2(x), \lambda), W(T_1(x), T_2(y), \lambda))$$
$$+ d(W(T_1(x), T_2(y), \lambda), W(T_1(y), T_2(y), \lambda))$$
$$\leq (1-\lambda)d(T_2(x), T_2(y)) + \lambda d(T_1(x), T_1(y))$$

(by properties (I^*) and (I))

$$\leq (1-\lambda) d(x, y) + \lambda d(x, y)$$

(as T_1, T_2 are non-expansive)

$$= d(x,y)$$

implying that T is a non-expansive mapping on C.

Now we show that $F(T) = F(T_1) \cap F(T_2)$. Let $x \in F(T_1) \cap F(T_2)$ i.e. $T_1(x) = x$ and $T_2(x) = x$. Then $T(x) = W(T_1(x), T_2(x), \lambda) = W(x, x, \lambda) = x$ as

$d(W(x,x,\lambda), x) \leq \lambda d(x, x) + (1-\lambda) d(x, x) = 0$. Therefore $x \in F(T)$ $F(T_1) \cap F(T_2) \subseteq F(T)$.

Now suppose $x \in F(T)$ and $y_o \in F(T_1) \cap F(T_2)$.

Consider

$$d(x, y_0) = d(Tx, y_0)$$

$$= W(T_1(x), T_2(y), \lambda), y_0)$$

$$\leq \lambda d(T_1(x), y_0) + (1-\lambda) d(T_2(x), y_0)$$

(by the convexity of X)

$$= \lambda \, d(T_1(x), T_1(y_0)) + (1-\lambda) \, d(T_2(x), T_2(y_0))$$

$$(\text{as } y_o \in F(T_1) \cap F(T_2))$$

$$\leq \lambda \, d(x, y_0) + (1\lambda) \, d(x, y_0)$$

$$(\text{as } T_1, T_2 \text{ are non-expansive})$$

$$= d(x, y_0).$$

Thus equality holds throughout and so

$$d(W(T_1(x), T_2(x), \lambda), y_0) = d(x, y_0).$$

Since $d(T_1(x), y_0) \leq d(x, y_0)$ and $d(T_2(x), y_0) \leq d(x, y_0)$,

the strict convexity of X implies $T_2(x) = T_1(x)$. Therefore

$$x = Tx$$
$$= W(T_1(x), T_2(x), \lambda)$$
$$= W(T_1(x), T_1(x), \lambda)$$
$$= T_1(x)$$

Thus $x = T_1(x) = T_2(x)$ and so $x \in F(T) = F(T_1) \cap F(T_2)$ implying

$F(T) \subseteq F(T_1) \cap F(T_2)$.

It was shown in [16] that for a closed convex subset C in a real, reflexive, strictly convex Banach space X, the class of non-expansive retracts of C is closed under arbitrary intersection. The same conclusion was drawn in [17] when C is a non empty closed, convex, locally weakly compact subset of a strictly convex Banach space X. In convex metric spaces we have:

Theorem 8.3 [81]. Suppose (X,d) is a complete strictly convex metric space with properties (I) and (I^*), C a locally compact, convex set in X and T_1, T_2: C→C be non-expansive mappings. Then F $(T_1)\cap$F (T₂) is a non-expansive retract of C and hence metrically convex.

Proof. If F $(T_1)\cap$F (T₂) $=\phi$, then clearly F $(T_1)\cap$F (T₂) is a non-expansive retract of C and thence by Theorem 6.1 is metrically convex.

Now suppose F (T_1) $\cap$F $(T_2)=\phi$ then by Lemma 6.4, there exists a non-expansive mapping T:C → C defined by

T(x) = W $(T_1, T_2$, λ) (x), 0≤ λ ≤ 1 such that f(T) = F $(T_1)\cap$F $(T_2$). We show that F $(T_1)\cap$F $(T_2$) = F (T) is a non-expansive retract of C.

Let x ∈F (T) = F $(T_1)\cap$F $(T_2$). Then T_1 (x) = x and T_2 (x) = x.

Consider

$$d(Tx, x) \quad = d(w(T_1\ (x)\ ,\ T_2\ (x)\ ,\lambda\ )\ ,x)$$
$$\leq \lambda d\ (T_1\ (x)\ ,\ x) + (1-\lambda\ )\ d(T_2\ (x)\ ,x))$$

(by the convexity of X)

$$= \lambda\ d\ (x,\ x) + (1-\lambda\ )\ d\ (x,\ x)$$
$$= 0$$

and so Tx = x i.e. x is a fixed point of T in C.

Let K be a bounded closed convex set in C such that T leaves K invariant.

Now $T^* = T/K : K \to K$ is non-expansive as T:C→C is a non-expansive mapping.

Since every convex set is starshaped with respect to each of its elements so is K . Let p be a starcentre of K and $T_p : K \to K$ be a mapping defined by $T_p(x) = p$ for all $x \in K$.

Let $< k_n >$ be any sequence of real numbers with $0 \leq k_n < 1$ and $k_n \to 1$.

Define $T' : K \to K$ by $T'(x) = W(T', T_p, k_n)(x)$ for all $x \in K$.

Consider

$$d(T'x, T'y) = d(W(T^*, T_p, k_n)(x), W(T^*, T_p, k_n)(x)$$

$$= d(W(T^*(x), p, k_n), W(T^*(y), p, k_n))$$

103

$$\leq k_n d(T^*(x), T^*(y)) \qquad\qquad \text{(by property (I))}$$

$$\leq k_n d(x, y) \qquad\qquad T(\quad \text{being non-expansive)}$$

Thus T' is a k_n-contraction on K and so by Banach contraction principle, T' has a fixed point, say x_1 in K i.e $T' x_1 = x_1$.

Now,

$$d(x_1, T' x_1) = d(T' x_1, T' x_1)$$

$$= d(W(T', T_p, k_n)(x_1), T'(x_1))$$

$$= d(W(T'(x_1), p, k_n), T'(x_1))$$

$$\leq k_n d(T^*(x_1), T^*(x_1)) + (1 - k_n) d(p, T^*(x_1)) \qquad \text{(by the convexity of X)}$$

$$\to 0 \text{ as } n \to \infty .$$

So for $x_1 \in K$, we have $T^*(x_1) = x_1$ i.e. $T^* = T/K$ has a fixed point in K implying that T has a fixed point x_1 in every

non-empty closed convex set K that T leaves invariant i.e. T satisfies (CFP).

Thus we get a non-expansive mapping $T:C \to C$ satisfying (CFP). So, by Theorem 8.2, F(T) is a non-expansive retract of C and hence by Theorem 8.1, $F(T_1) \cap F(T_2) = F(T)$ is metrically convex.

--- X ---

REFERENCES

[1] G.C. Ahuja, T.D. Narang and Swaran Trehan : Best Appoximation on Convex Sets in Metric Linear   Spaces, Math. Nachr., 78(1977), 125-130.

[2]G.C. Ahuja and T.D. Narang :   On   Best   Simultaneous   Approximation, Nieuw Arch. Wisk., 27(1979), 255-261.

[3]G.C. Ahuja, T.D. Narang and Swaran Trehan: Best Approximation on Convex Sets in a Metric Space, J. Approx. Theory, 12 (1974), 94-97.

[4]G. Albinus :Uber   Bestapproximationen   in   metrischen   Vektorraumen, Dissertation, Tech.   Univ.   Dresden (1966).

[5]G. Albinus   : Einige Beitrage zur Approximations - theorie in metrischen Vektorraumen, Wiss. Z.d.Tech. Univ. Dresden, 15(1966), 1- 4.

[6]G.Albinus: Normartige   Metriken   auf   metrisierbaren   lokalkonexen topologischen Vaktoraumen, math.Nachr. 37(1968) , 177-196.

[7]I. Beg and A. Azam :   Fixed Points of Multi- valued Locally Contractive Mappings,   Boll. U.M.I., 4-A   (1990), 227-233.

[8]Ismat Beg,   Naseer Shahzad and Mohammd Iqbal   : Fixed Point Theorems and Best Approximation in Convex Metric Spaces, Approx. Theory & its Appl. , 8 (1992), 97-105.

[9]Ismat Beg and Naseer Shahzad : An Application   of   a   Fixed Point Theorem to Best Simultaneous   Approximation,   Approx. Theory & Its Appl., 10 (1994), 1-4.

[10]L.P. Belluce and W.A. Kirk : Fixed-Point   Theorems   for families of Contraction Mappings, Pacific J.   Math, 18 (1966), 213-217.

[11]G.Birkhoff : Orthogonality in Linear Metric Spaces, Duke Math. J.1 (1935) , 169-172.

[12]S.C.   Bose :   Introduction to Functional Analysis ,   Macmillan India Limited (1992).

[13] N. Bourbaki , Topological Vector spaces, Addison – Wesley (1955) .

[14]B. Brosowski :   Fixpunktsatze   in der   Approximation- theoris, mathematica (Cluj),   11(1969), 195-200.

[15] A.L. Brown : Abstract Approxmation Theory, Seminar   in Analysis, Matscience, Institute   of Mathematical   Sciences, Madras (1969-70).

[16]Ronald E. Bruck Jr. : Non-expansive Retracts in Banach   Spaces, Bull. Amer. Math. Soc., 76(1970), 384-386.

[17]Ronald E. Bruck. Jr. : Properties of Fixed Point sets   of Non-expansive Mappings in Banach Spaces Trans.   Amer. Math. Soc., 179(1973), 251-262.

[18] R.C. Buck : Applications of Duality in Approximation   Theory - Approximation of Functions,   (Ed. H.L.   Garabedian) Elsevier Publ. Co. , Amsterdam – London   - New York, (1965), 27-42.

[19]E.W.Cheney : Introduction to Approximation Theory,   McGraw Hill, New York (1966).

[20]James A. Clarkson :   Uniformly Convex Spaces, Trans. Amer. Math. Soc., 40 (1936), 396-414.

[21]Philip J. Davis : Interpolation and Approximation,   Blaisdell Publishing Company, New York (1963).

[22]J.B. Diaz and H.W. Mchaughlin : on   Simultaneous   Chbyshev Approximation and Chebyshev Approximation   with an Additive Weight, J.Approx. Theory, 6(1972),   68-71.

[23]Sever Silvestru Dragomir : on Approximation of   Continuous Linear Functionals in Normed Linear   Spaces, Analele Universitatii din   Timisora, 29(1991), 51-58.

[24]C.B. Dunham :   Simltaneous Chebyshev   Approximation of Functions on an Interval, Proc. Amer.   Math. Soc., 18(1967),   472-477.

[25] N.V. Efimov and S.B. Steckin : Approximative Compact - ness and Chebyhev Sets, Dokl. Akad. Nauk, SSSR, 140(1961), 522-524.

[26]C. Franchetti and M.Furi : Some Properties of Hilbert Spaces, Rev. Roum. Math. Pure et Appl., 17 (1972), 1045-1048.

[27]D.S. Goel, A.S.B. Holland, C. Nasim and B.N. Sahney : On Best Simultaneous Approximation in Normed Linear Spaces, Canad. Math. Bull. , 17 (1974), 523- 527.

[28]Paladugu Govindrajulu : On Best Simultaneous Approximation , J. Math. Phy. Sci.,18 (1984), 345- 351.

[29]L.F. Jr. Guseman and B. C. Perers : Non expansive Mappings on Compact Subsets of Metric Linear Spaces, Proc. Amer. Math Soc., 47(1975), 383-396.

[30]E. Hewitt and K.Stromberg : Real and Abstract Analysis, Springer-Verlag, Heidelberg-Berlin (1968).

[31]T.L. Hicks and M.D. Humphries : A note on Fixed Point Theorems, J. Approx. Theory, 34(1982), 221-225.

[32] D.Hinrichsen and H. Bauer: Einige Eigenscharten Lokelkompakter konvexer mengen. Undiherer projektiven limited , in proc. Collaq. Convexity, Coperhagen, 1965 , Kobenhavns Universistets Matematiske Institut, Copenhagen (1967), 143-153

[33] R.A. Hirschfeld : On Best Approximation in Normed Vector Spaces, Nieuw Arch. Wisk., 6(1958), 41-51.

[34].R.A. Hirschfeld: On Approximation Theory, Birkhauser Verlag (1964).

[35]R.B. Holmes : A course on optimization and Best Approximation, Lecture Notes , Springer Verlag (1972) .

[36]Vasile I. Istratescu : Fixed Point Theory, D. Reidal Publishing Company Dordrecht, Holland (1981).

[37] Vasile I. Istratescu : Strict Convexity and Complex Strict Convexity, Theory and Applications , Marcel Dekker, Inc. New York (1984).

[38]Paul C. Kainen, Vera Kurkova and Andrew Yogy: Geometry and Topology of Continuous Best and Near Best Approximation, J. Approx. Theory, 105(2000), 252-262.

[39]L.A. Khan and A.R. Khan : An Extension of Brosowski- Meinardus Theorem on Invariant Approximation, Approx. Theory & its Appl., 11(1995), 1-5.

[40]V.L.Klee : Convex Bodies and Periodic Homeomorphism in Hilbert Spaces, Trans. Amer. Math. Soc., 74(1953), 10-43.

[41]G.Kothe: Topological Vector Spaces I , Springer- Verlag (1969).

[42] I.M. Liberman: some Characterstic properties of convex bodies , Mat. Sb., 13(1943), 239-262.

[43]G. Meinardus : Invarianz bei linearen, Appriximation, Arch. Rational Mech. Anal., 14(1963), 301-303.

[44]K. Menger : Untersuchungen uber allgemeine Metrik, Math. Ann., 100(1928), 75-163.

[45]R.N. Mukherjee and V. Verma : Some Fixed Point Theorems and Their Applications to Best Simultaneous Approximation, Publications de L'Institut Mathematique, Nouvelle Serie Tome, 49(63) (1991), 111-116.

[46]T.D. Narang: Study of Nearest and Farthest Points on Convex Sets, Ph.D. Thesis, University of Delhi (1975).

[47]T.D. Narang : On Certain Characterizations of Best Approximations in Metric Linear Spaces, Pure and Applied Mathematica Sciences, 4(1976), 1-2.

[48]T.D. Narang: on certain Characterizations of Best Approximation in metric linear spaces, Pure and Applied Mathematika Sciences 4 (1976), 121-124.

[49]T.D. Narang : Best Approximation and Strict Convexity of Metric Spaces, Arch. Math. Math., 17(1981), 87-90.

[50]T.D. Narang : On Totally Complete Spaces,The Mathematics Education, 16(1982), 4-5.

[51]T.D. Narang: Unicity Theorem and Strict Concvexity of or metric linear spaces, Tamakang J.Math. 11(1980).49-51, Corrigendum, 14(1983), 103-104

[52]T.D. Narang : On Best Simultaneous Approximation, J. Approx. Theory, 39(1983), 93-96.

[53]T.D. Narang : Applications of Fixed Point Theorems to Approximation Theory, Mat. Vesnik, 36(1984), 69-75.

[54]T.D. Narang: A characterization or Strictly convex Metrix linear Space, 39 (1986), 149-151.

[55].T.D. Narang : on Best Co-approximation in Normed Linear Spaces, Rocky Mountain Journal of Mathematics, 22(1991), 265-287.

[56]T.D. Narang: Simultaneous Approximation and Chebyshev Centers in Metric Spaces, Mat. Vesnik, 51(1999), 61- 68.

[57]T.D. Narang : On Best Simultaneous Approximation in Metric Spaces, To Appear in Universitatii din Timisoara Seria Matematica –Informatica (Romania)

[58]Neera: On strictly Convex Normed Linear Spaces. M.Phil. Dissertation, G.N.D. University (1985)

[59]R.R. Phelps : Convex Sets and Nearest Points, Amer. Math. Soc., 8(1957), 790-797.

[60]R.R. Phelps: Convex sets and nearest points, Proc. Amer. Math. Sec. 8 (1975), 790-797

[61]M.J.D. Powel : Approximation Theory and Methods, Cambridge University, Cambridge (1981)

[62]G.S. Rao : Best Co-Approximation in Normed Linear Spaces, Approximation Theory V. C.K. Chui, L.L. Schumaker and J.D. Ward (Eds.), Academic Press, New York (1986), 535-538.

[63]Geetha S.Rao: Approximation Theory and its Applications, New Age International, New Delhi (1996).

[64]Geetha S.Rao and R. Sarvanan: Best Simultaneous Co- approximation, Indian J. Math., 40(1998), 353-362.

[65]Walter Rudin: Functional Analysis, Tata Mc Graw-Hill (1973)

[66]B.N. Sahney and S.P. Singh : On Best Simultaneous Chebyshev Approximation with Additive Weight Functions and Related Results, Nonlinear Analysis and Applications, Lecture Notes in Pure and Applied Mathematics (Eds. S.P. Singh and J.H. Burry) Marce I Dekker, Inc. (1982), 443-463.

[67]K.P.R. Sastry and S.V.R. Naidu: Convexity Conditions in Metric Linear Spaces, Mathematics Seminar Notes 7 (1979), 235-251

[68]K.P.R. Sastry and S.V.R. Naidu : Upper Semi-Continuity of Best Simultaneous Approximation Operator, Pure and Appl. Math. Sci., 10(1979),7-8.

[69]K.P.R. Sastry and S.V.R. Naidu : Uniform Convexity and Strict Convexity in Metric Linear Spaces, Math. Nachr., 104(1981), 331-347.

[70] K.P.R. Sastry and S.V.R. Naidu: Proximity and Ball Convexity in Metric Linear Spaces, Tamkang J. Math 14 (1983) 5-7.

[71]K.P.R. Sastry, S.V.R. Naidu and M.V.K. Ravi Kishore : Pseudo Strict Convexity and Metric Convexity in Metric Linear Spaces, Indian J. Pure Appl. Math., 19(1988), 149-153.

[72]Ioan Serb: on the Multi-Valued Metric Projection in Normed Vector Spaces, Rev Anal. Numer. Theory Approx., 10(1981), 101-111.

[73] Ioan Serb: on the Multi-Valued Metric Projections in Normed Vector Spaces II, Anal. Numer. Theory Approx., 10(1982), 155-166.

[74]Ioan Serb: Strongly Proximinal Sets in Abstract Spaces, Seminar on Functional Analysis and Numerical Methods, Preprint Nr. 1(1984), 159-168.

[75] Meenu Sharma and T.D. Narang : On ε-Birkhoff Orthogonality and ε -Near Best Approximation , Journal of the Korea Society of Mathematical Education, 8(2001), 153-162.

[76]Meenu Sharma and T.D.Narang: On ε- Simultaneous Approximation J.Nat.Acad.Math.,16(2002),pp.7-14.

[77]Meenu Sharma and T.D.Narang:On ε- Simultaneous Approximation and

A Fixed Point Theorem – J.Nat.Acad.Math.,16(2002),pp.29-36.

[78]Meenu Sharma and T.D. Narang : On Best Approximation and Fixed Points in Pseudo Strictly Convex Spaces, Current Trends in Industrial and Applied Mathematics, P. Manchanda, K. Ahmed and A.H. Siddiqui (Eds.), Anamaya Publishers, New Delhi (2002).

[79]Meenu Sharma and T.D.Narang:On Invariant Approximation of Non-Expansive Mappings- -- Journal of Analele Universitatii din Timisoara of Romania, 41(2003), 117-127

[80]Meenu Sharma and T.D.Narang:On The Multivalued Metric Projections in Convex Spaces --Journal of Analele Universitatii din Timisoara of Romania, 41(2003), 143-152.

[81]Meenu Sharma and T.D. Narang: On Non-Expansive Retracts in Convex Metric Spaces-Journal of the Korea Society of Mathematical Education,10(2003), 127-132.

[82] Meenu Sharma and T.D.Narang: On Best Simultaneous Co-approximation- "Mathematics in the 21st Century" by Prof. K.K. Dewan and M.Mustafa (Editions) Deep & Deep Publications, Delhi (2004).

[83].Horald S. Shapiro : Topics in Approximation Theory, Springer – Verlag, New York(1971).

[84].Ivan Singer : Best Approximation in Normed Linear Spaces by Elements of Linear Subspaces, Springer- Verlag, New York(1970)

[85]Ivan Singer : Best Approximation in Normed Spaces, Constructive Aspects of Functional Analysis – II Parte, Ediziona Cremonese, Roma (1973).

[86].S.P. Singh : An Application of a Fixed Point Theorem to Approximation Theory, J. Approximation Theory, 25 (1979), 89-90.

[87]S.P. Singh : Application of Fixed Point Theorem in Approximation Theory, Applied Non-Linear Analysis (Ed. V. Lakshmikanthan) Academic Press, Inc. New York (1979),389-394.

[88]D.R. Smart : Fixed Point Theorems, Cambridge Univ. Press, Cambridge, U.K. (1974).

[89]S.B. Steckin : Approximation Properties of Sets in Normed Linear Spaces (Russian), Rev. Roumaine Math. Pure. Appl., 8 (1963), 5-18.

[90]W. Takahashi : A Convexity in Metric Spaces and Non- Expansive Mappings I, Kodai Math. Sem. Rep., 22 (1970), 142-149.

[91] M.A. Al-Thagafi : Best Approximation and Fixed Points in Strong M-Starshaped Metric Spaces, Internat. J. Math. And Math. Sci., 18(1995), 613-616.

[92] A.I. Vasil'ev: The Structure of Balls and Uniqueness Subspaces in a Metric Linear Space. Mat. Zametki, 13 (1973), 541-550

[93] A.I. Vasil'ev: Chebyshev Sets and Strong Convexity or Metric Linear Spaces, Math. Notes 25 (1979), 335-340.

[94]. L.P. Vlasov : Approximative Properties of Sets in Normed Linear Spaces, Russian Math. Surveys, 28 (1973),1-66.

--- X ---

Printed in the United States
By Bookmasters